Marcello Stein

Characterization of the CD83 promoter/enhancer complex

Marcello Stein

Characterization of the CD83 promoter/enhancer complex

in human dendritic cells

Südwestdeutscher Verlag für Hochschulschriften

Impressum / Imprint
Bibliografische Information der Deutschen Nationalbibliothek: Die Deutsche Nationalbibliothek verzeichnet diese Publikation in der Deutschen Nationalbibliografie; detaillierte bibliografische Daten sind im Internet über http://dnb.d-nb.de abrufbar.
Alle in diesem Buch genannten Marken und Produktnamen unterliegen warenzeichen-, marken- oder patentrechtlichem Schutz bzw. sind Warenzeichen oder eingetragene Warenzeichen der jeweiligen Inhaber. Die Wiedergabe von Marken, Produktnamen, Gebrauchsnamen, Handelsnamen, Warenbezeichnungen u.s.w. in diesem Werk berechtigt auch ohne besondere Kennzeichnung nicht zu der Annahme, dass solche Namen im Sinne der Warenzeichen- und Markenschutzgesetzgebung als frei zu betrachten wären und daher von jedermann benutzt werden dürften.

Bibliographic information published by the Deutsche Nationalbibliothek: The Deutsche Nationalbibliothek lists this publication in the Deutsche Nationalbibliografie; detailed bibliographic data are available in the Internet at http://dnb.d-nb.de.
Any brand names and product names mentioned in this book are subject to trademark, brand or patent protection and are trademarks or registered trademarks of their respective holders. The use of brand names, product names, common names, trade names, product descriptions etc. even without a particular marking in this works is in no way to be construed to mean that such names may be regarded as unrestricted in respect of trademark and brand protection legislation and could thus be used by anyone.

Coverbild / Cover image: www.ingimage.com

Verlag / Publisher:
Südwestdeutscher Verlag für Hochschulschriften
ist ein Imprint der / is a trademark of
AV Akademikerverlag GmbH & Co. KG
Heinrich-Böcking-Str. 6-8, 66121 Saarbrücken, Deutschland / Germany
Email: info@svh-verlag.de

Herstellung: siehe letzte Seite /
Printed at: see last page
ISBN: 978-3-8381-3668-4

Zugl. / Approved by: Erlangen, FAU, Diss., 2011

Copyright © 2013 AV Akademikerverlag GmbH & Co. KG
Alle Rechte vorbehalten. / All rights reserved. Saarbrücken 2013

Part of the experimental data was published in the scientific journal *"Molecular and Cellular Biology"* as scientific article with the title *"Multiple IRF- and NFkB-sites cooperate in mediating cell type- and maturation-specific activation of the human CD83 promoter in dendritic cells"* (April 2013, volume 33, issue 7. Mol. Cell. Biol. 2013, 33(7):1331. DOI: 10.1128/MCB.01051-12. Published online ahead of print on 22 January 2013, MCB Accepts. Copyright © 2013, American Society for Microbiology. All Rights Reserved.)

Table of contents

Abbreviations and Symbols

1. Summary — 17

2. Introduction — 19

2.1.	The biology of dendritic cells	20
2.1.1.	DC subsets	20
2.1.2.	DCs as immunological sentinels	23
2.1.2.1	Antigen capture	23
2.1.2.2	Antigen processing	24
2.1.2.3.	Activation of DCs	25
2.1.2.3.1.	Innate immunity activation signals	25
2.1.2.3.2.	Adaptive immunity activation signals	28
2.1.2.4.	Maturation of DCs	28
2.1.2.5.	Migration of DCs	29
2.1.2.6.	Control of different T cell responses by DCs	29
2.1.3.	The surface molecule CD83	31
2.1.3.1.	Structure of CD83 gene and protein	32
2.1.3.2.	Functions of membrane bound and soluble forms of CD83	33
2.1.3.3.	CD83 as a target for viral immune escape mechanisms	35
2.1.3.4.	The CD83 promoter	35
2.1.4.	Adenoviruses and gene therapy	36
2.2.	Genetic regulation and expression mechanisms	38
2.2.1.	Structural modifications of DNA	39
2.2.2.	DNA Methylation	39
2.2.3.	Histone modification and chromatin remodeling	41
2.2.4.	Histone methylation	42
2.2.5.	Histone acetylation	43
2.2.6.	Regulation of transcription	43
2.2.7.	Regulatory DNA elements	43
2.2.7.1.	Promoters	45

2.2.7.2.	Upstream regulatory elements	46
2.2.7.3.	Enhancers	47
2.2.7.4.	Silencers	48
2.2.7.5.	Insulators	48
2.2.8.	DNA binding proteins	49
2.2.8.1.	The role of transcription factors during regulation of transcription	49
2.2.9.	The transcription factor SP1	50
2.2.10.	NFκB-family of transcription factors	51
2.2.10.1.	Regulation of NFκB	52
2.2.10.2.	NFκB-pathways	53
2.2.10.3.	The role of NFκB-transcription factors in the development of DCs	55
2.2.11.	Interferon regulatory factors	56
2.2.11.1.	Activation and regulation of IRF-pathways	57
2.2.11.2.	IRFs in development and function of DCs	58
2.2.11.3.	The role of IRFs in DC activation	61

3. Tasks 65

4. Material and Methods 67

4.1.	**Material**	67
4.1.1.	Chemicals	67
4.1.2.	DNA modifying enzymes	68
4.1.3.	Buffers and reagents	69
4.1.3.1.	General reagents and buffers	69
4.1.3.2.	Buffers and reagents for DEAE-Dextran cell transfection	69
4.1.3.3.	Buffer and reagents for SDS- polyacrylamide gel electrophoresis (PAGE) and Western blot analyses	70
4.1.3.4.	Reagents for chemical-competent E.coli and transformation by heat	71
4.1.3.5.	Buffers and reagents for electrophoretic mobility shift assay (EMSA)	71

4.1.3.6.	Buffers and reagents for DNA electrophoresis	72
4.1.3.7.	Ladders and markers	72
4.1.4.	Cell culture	73
4.1.4.1.	Cell lines and cell culture media	73
4.1.4.2.	Additional cell culture media	74
4.1.4.3.	General cell culture reagents	75
4.1.4.4.	General cell culture dishes and plastic ware	75
4.1.5	Equipment and instruments	75
4.1.6.	Software and websites	77
4.1.7.	Antibodies	77
4.1.7.1.	Antibodies used for FACS	77
4.1.7.2.	Antibodies used for EMSAs	78
4.1.7.3.	Antibodies used for Western blot analyses	79
4.1.8.	DNA Oligos	80
4.1.8.1.	Primers	80
4.1.8.1.1.	Sequencing primers	80
4.1.8.1.2.	Cloning primers for transcription factors	81
4.1.8.1.3.	Cloning primers for luciferase reporter constructs	81
4.1.8.1.4.	Cloning primers for the mutated 185 bp enhancer	82
4.1.8.2.	Oligos for EMSA	83
4.1.8.2.1.	Oligos NFκB-sites	83
4.1.8.2.2.	Oligos IRF-sites	83
4.1.9.	Plasmid vectors	84
4.1.9.1.	Donated or purchased vector	84
4.1.9.2.	Cloned vectors	85
4.1.9.2.1.	Luciferase reporter constructs	85
4.1.9.2.2.	Luciferase reporter constructs with mutated IRF sites	87
4.1.9.2.3.	Schematic depiction of CD83 Intron 2 fragments for luciferase assay	89
4.1.10.	Adenoviruses	90
4.1.11.	Human cytokines and maturation agents	90
4.1.12.	Bacteria	91
4.1.13.	Purchased Kits	91

4.2.	**Methods**	92
4.2.1.	General molecular biology methods	92
4.2.1.1.	Generation of chemical-competent *E.coli* and transformation by heat	92
4.2.2.	DNA based molecular biology methods	93
4.2.2.1.	Isolation of DNA	93
4.2.2.1.1.	Isolation of small plasmids (< 12 kb)	93
4.2.2.1.2.	Isolation of large plasmids (> 12 kb)	93
4.2.2.1.3.	Isolation of human genomic DNA from cells by QIAamp DNA Mini Kit	94
4.2.2.1.4.	Preparation of DNA fragments from PCR reactions and enzymatic digestions	94
4.2.2.1.5.	Preparation of DNA fragments from gel electrophoresis	94
4.2.2.2.	Nucleic acid quantification	95
4.2.2.3.	Separation of DNA or RNA using agarose gel electrophoresis	95
4.2.2.4.	Polymerase chain reaction (PCR) based methods	95
4.2.2.4.1.	Single-step PCRs	95
4.2.2.5.	Hybridization of DNA oligos for elctrophoretic mobility shift assay (EMSA)	96
4.2.2.6.	Radioactive labeling of DNA oligos for EMSA	97
4.2.2.7.	Cloning of DNA fragments or PCR products	98
4.2.2.7.1.	Dephosphorylation of cleaved DNA	98
4.2.2.7.2.	Conversion of DNA overhangs	98
4.2.2.7.3.	Ligation	99
4.2.2.7.4.	Sequencing of DNA	99
4.2.2.7.5.	Synthesis of DNA sequences	99
4.2.3.	Protein biochemistry methods	99
4.2.3.1.	Generation of cell lysates	99
4.2.3.1.1.	Preparation of whole cell extracts for SDS-polyacrylamide gel electrophoresis (PAGE)	99
4.2.3.1.2.	Preparation of nuclear extracts for EMSA and SDS-PAGE	100
4.2.3.1.3.	BCA Protein Assay Reagent (bicinchoninic acid)	100
4.2.3.2.	EMSA	101
4.2.3.2.1.	EMSA bandshift and supershift reaction	101

4.2.3.2.2.	EMSA non denaturizing polyacrylamide gel run and signal detection	102
4.2.3.3.	SDS-PAGE: Laemmli method	103
4.2.3.4.	Western blot analyses	103
4.2.4.	Cell culture	104
4.2.4.1.	Generation of dendritic cells (DC)	104
4.2.4.2.	Cryopreservation of primary cells	105
4.2.4.3.	Cryopreservation of cell lines	105
4.2.4.4.	Heat inactivation of fetal calf (FCS) and human autologous serum	105
4.2.5.	Flow cytometric analysis (FACS)	105
4.2.6.	Transient transfection methods and luciferase reporter assay	106
4.2.6.1.	DNA-transfection using the DEAE-Dextran method	106
4.2.6.2.	Electroporation of Raji and Jurkat cell lines	106
4.2.6.3.	Lipofection of DNA with Lipofectamine^{TM} LTX and PLUS^{TM} reagent	107
4.2.6.4.	Electroporation of DCs with DNA using AMAXA technology	108
4.2.6.5.	Luciferase reporter assay	109
4.2.7.	Recombinant adenoviruses	109
4.2.7.1.	Cloning of plasmids containing the recombinant adenoviral genome	109
4.2.7.2.	Preparation of recombinant adenoviruses	109
4.2.7.3.	Determination of the physical particle concentration	110
4.2.7.4.	Determination of the infectious particle concentration	111
4.2.7.5.	Adenoviral transduction of cells	111
4.3.	Statistical analysis	112

5. Results 113

5.1.	Previous approaches to identify the human CD83 promoter	113
5.1.1.	CD83 is not expressed in HFF cells, weakly in iDCs and strongly in mDCs	113
5.1.2.	A 6 kB region of CD83 intron 2 is specifically H3K9 acetylated in mDCs	115

5.2.	Analyses of the acetylated region of intron 2 within the CD83 gene by luciferase reporter assays	118
5.2.1	CD83 Intron 2 Fragment C shows an enhancing activity on MP -261 in the DC-like cell lineXS52	120
5.2.2.	Narrowing down the putative enhancer sequence: A 185 bp sequence within fragment C shows enhancer function	122
5.2.3.	The CD83 intron 2 fragment C and the *185 bp enhancer* induce the MP -261 specifically in human DCs	125
5.2.4.	CD83 intron 2 fragment C from and the *185 bp enhancer* do not induce the MP -261 in B and T cell lines	127
5.3.	Biocomputational analyses and modeling of the human CD83 promoter	129
5.4.	Functional characterization of the predicted UpP and the spacer sequence S1 within the pGL3/MP -261 reporter plasmids	131
5.4.1.	Neither the spacer sequence S1, nor the UpP do significantly affect the induction of the CD83 promoter in cell lines	132
5.4.2.	The spacer sequence S1 does not significantly affect the induction of the CD83 promoter in mDCs.	134
5.5.	Adenoviral transduction confirms the cell type- and status-specific interaction of UpP, MP -261 and *185 bp enhancer*	136
5.5.1.	The proposed ternary complex of UpP, MP -261 and the *185 bp enhancer* forms in mDCs	137
5.5.2.	The proposed ternary complex does not form in Raji, Jurkat and JCAM cell lines	139
5.6.	Analyses of the predicted transcription factor binding sites by electrophoretic mobility shift assays (EMSA)	142
5.6.1.	Bandshift analyses of the predicted NFκB- and IRF-sites	142
5.6.2.	Supershift analyses for the transcription factors of the NFκB-family	144
5.6.2.1.	The statistical evaluation confirms the results of the EMSAs for the NFκB- sites	150
5.6.3.	Supershift analyses for the transcription factors of the IRF-family	153
5.6.3.1.	The statistical evaluation confirms the results of the EMSAs for the IRF-sites	156
5.6.4.	Summary of the EMSAs	158
5.7.	Functional analyses of the three IRF-sites using mutagenesis and luciferase assays	160

5.7.1.	Mutation of individual IRF-sites results in a significantly reduced luciferase expression in XS52 cells and mDCs	161
5.8.	Verification of the functionality of the NFκB-sites in the UpP and the MP -261 by cotransfection experiments in 293T cells	165
5.8.1.	IRF-5, p50, p65 and cRel can be overexpressed in 293T cells	166
5.8.2.	IRF-5, p50, p65 and cRel induce the MP -261 in 293T cells	167
5.8.3.	The combination of p65 and IRF-5 induces the UpP in 293T cells	170
5.9.	The transcription factors IRF-1, IRF-2, IRF-5, p50, p65 and cRel are differentially expressed in iDCs, mDCs and HFF cells	172
5.10.	The DNA in the CD83 promoter region is not differentially CpG methylated in iDCs, mDCs and HFF cells	175

6. Discussion 179

6.1.	Full characterization of the human CD83 promoter	179
6.2.	A 185 bp sequence within CD83 intron 2 has mDC-specific enhancer function	179
6.3.	A ternary complex consisting of MP -261, *185 bp enhancer* and UpP regulates mDC-specific CD83 expression	181
6.4.	The formation of the ternary complex is mediated by NFκB- and IRF-transcription factors	185
6.4.1.	Verification of the NFκB-transcription factor binding sites	186
6.4.2.	Verification of the IRF-transcription factor binding sites	190
6.5.	Summary and future prospects	192

7. References 196

Acknowledgements (German/Italian)

Patents, Publications, Presentations, Participations and Commitments

Abbreviations and Symbols

µg	Microgram	mCD83	Membrane bound CD83
µl	Microliter	SP1	Specificity factor 1
Ad	Adenovirus	mg	Milligram
APC	Antigen presenting cell	L-DC	Lymphoid dendritic cell
as	Antisense	n.s.	Not significant
bp	Basepair	TFBS	Transcription factor binding site
CDS	Coding sequence	HAT	Histone acetyltransferase
ChIP	Chromatin immuno-precipitation	IL	Interleukine
CTL	Cytotoxic T cell	PRR	Pattern recognition receptor
d	Day	vp	Virus particle
DC	Dendritic cell	LPS	Lipopolysaccharide
E	Exon	BS	Binding site
E.as	Enhancer in antisense orientaion	°C	Degree Celsius
E.s	Enhancer in sense orientaion	RT	Room temperature
EMSA	Electrophoretic mobility shift assay	RNA	Ribonucleic acid
Enh.	Enhancer	cpm	Counts per minute
FACS	Fluorescence assisted cell sorting	RT-PCR	Real time polymerase chain reaction
Fig.	Figure	Treg	Regulatory T cell
h	Hour	TLR	Toll like receptor
H3K9	Acetylation of lysine 9 of histone 3	GM-CSF	Granulocyte macrophage colony-stimulating factor
HFF	Human foreskin fibroblast	PCR	Polymerase chain reaction
HSV	Herpex simplex virus	HCMV	Human Cytomegalovirus

I	Intron	Ac	Acetylation
iDC	Immature dendritic cell	$TCID_{50}$	Tissue culture infectious dose 50
IFN	Interferon	LC	Langerhans cell
intDC	Interstitial dendritic cell	cDC	Conventional dendritic cell
IRF	Interferon regulatory factor	mRNA	Messenger ribonucleic acid
kDa	Kilodalton	MHC	Major histocompatibility complex
mAb	Monoclonal antibody	OD	Optical density
MC	Maturation cocktail	HDAC	Histone deacetylase
mDC	Mature dendritic cell	PBMC	Perinuclear
M-DC	Myeloid dendritic cell	P-DC	Plasmacytoid dendritic cell
min	Minute	sCD83	soluble CD83
MP -261	261 basepair long CD83 minimalpromoter	DNA	Deoxyribonucleic acid
MT	Mutant	rpm	Rounds per minute
mut	Mutated	ng	Nanogram
NFκB	Nuclear factor kappa B	aa	Amino acid
P -510	510 basepair long CD83 promoter fragment including the upstream promoter	e.g.	Exampla gratia (for example)
PAMP	Pathogen associated molecular pattern	L	Ligand
PGE_2	Prostaglandine E_2	NAF	Non adherent fraction
RBI	Relative band intensity	DEAE	Diethylaminoethyl cellulose
RLU	Relative light unit	PI	Propidium Iodide
s	sense	n	Number
S1	Spacer sequence 1	g	Gram
SEM	Standard error of mean	3C	Chromosome conformation capture
SV40	Simian virus 40	Ab	Antibody

TF	Transcription factor	**K**	Lysine
TGF	Transforming growth factor	**PBMC**	Peripheral blood mononuclear cell
Th	T helper cell	**PAGE**	Polyacrylamide gel electrophoresis
TNF	Tumor necrosis factor	**moDC**	Monocyte derived dendritic cells
UpP	Upstream promoter	**Ig**	Immune globulin
WT	Wildtype	**kb**	Kilobase
α	Alpha	**β**	Beta
γ	Gamma		

1. Summary

CD83 is the best known maturation surface marker for dendritic cells (DC). Being specifically upregulated upon DC maturation suggests a highly cell type- and maturation status-specific regulation of the CD83 expression. However, so far only the CD83 minimal promoter (MP -261) has been characterized. Despite being active in several cell types, the MP -261 showed no cell type- or maturation status-specific activity. Therefore, it was concluded that additional regulatory elements like cell type-specific enhancers provide for the tightly regulated CD83 expression in DCs. Thus, the aim of this work was to fully characterize the CD83 promoter as well as the underlying molecular mechanisms of the cell type- and maturation status-specific CD83 regulation.

The first approach to identify regulatory elements contributing to the cell type specificity of the CD83 expression was a ChIP-chip^{TM} microarray (in cooperation with Dr. I. Knippertz[422]) directed against acetylated lysine 9 of Histone 3 (H3K9) in immature DCs (iDC), mature DCs (mDC) and human foreskin fibroblasts (HFF). The ChIP-assay revealed a genomic region within CD83 intron 2 that was specifically H3K9 acetylated in mDCs, but not in iDCs and HFF cells.

Deletion mutagenesis and luciferase reporter assays elucidated a 185 bp long enhancer (*185 bp enhancer*) within this acetylated region that specifically induced the MP -261 in the DC-like cell line XS 52 and mDCs, whereas the induction was absent in the control cell lines as well as in iDCs.

A biocomputational analysis of the MP -261 in combination with the *185 bp enhancer* predicted three NFκB- and five SP1-sites in the MP -261 as well as two IRF-sites and one SP1-site in the *185 bp enhancer*. Furthermore, a third regulatory element, the CD83 upstream promoter (UpP), was proposed. Two additional NFκB-, one IRF- and one SP1-site were predicted to lie within the UpP. Furthermore, the biocomputational model predicted the interaction of those three regulatory elements to form three copies of a well known NFκB-IRF-NFκB transcriptional module *in trans*. The formation of such a module *in trans* represents a completely new molecular mechanism that has not been described so far.

Summary

To prove this model, a series of experiments was performed. First, adenoviral transduction of luciferase reporter vectors proved in a chromosome-like configuration that all three regulatory elements, namely UpP, MP -261 and *185 bp enhancer* have to be present in the same vector to induce transcription specifically in mDCs. This clearly proved the cooperation of all three regulatory elements to induce transcriptional activity *in vitro*.

Second, the ability to bind nuclear factors specifically from mDCs was proven by electrophoretic mobility shift assays (EMSA) for all predicted NFκB- and IRF-sites. It could be shown that IRF-5 binds to the IRF-site in the UpP, whereas p50 and cRel bind to the NFκB-sites in the MP -261. Moreover, IRF-1 and 2 were shown to bind to the proximal IRF-site in the *185 bp enhancer*. Although the remaining distal IRF-site in the enhancer and the two NFκB-sites in the UpP have been shown to bind nuclear factors from mDCs, the nature of these factors could not be revealed by EMSA.

Third, the function of the predicted IRF- and NFκB-sites has been verified individually by loss of function and gain of function experiments, respectively. On the one hand mutation of the IRF-sites in the UpP and the enhancer revealed that all three sites are necessary for the induction of transcription in the tripartite complex in XS52 cells and mDCs. On the other hand cotransfection of p50, p65 and cRel in combination with IRF-5 verified the functionality of the NFκB-sites by inducing luciferase reporter plasmids containing the UpP and the MP -261 in 293T cells.

Finally, a CpG-methylation analysis in iDCs, mDCs and HFF cells revealed that the CD83 promoter is not subjected to a cell type-specific epigenetic regulation.

Taken together, the cell type- and maturation status-specific expression of CD83 has been shown to be regulated by a tripartite complex consisting of UpP, MP -261 and *185 bp enhancer*. This complex forms specifically in mDCs and is mediated through the interaction of IRF- and NFκB-transcription factors.

2. Introduction

Dendritic cells (DCs) are the most potent antigen presenting cells (APC) of the mammalian immune system. They function as messengers between the innate and the adaptive immune system and are essential mediators of immunity and tolerance[1;2]. Among other APCs, namely B cells, $\gamma\delta$ T cells, monocytes and macrophages, DCs possess the unique ability to induce primary immune responses[3;4]. They capture and process antigens in the periphery, thereby inducing the primary immune response. Furthermore, they determine the type of T cell-mediated immune response and are able to induce immunological tolerance. Due to their central role in immunity DCs have attracted much attention in efforts to understand on the one hand the complex regulation of immunity/tolerance and on the other hand to develop DC-based vaccination strategies.

In this respect the cell surface marker CD83 is of high interest. CD83 is the best known maturation surface marker for DCs and is strongly upregulated during DC maturation exhibiting a tight regulation of expression in a cell type and maturation stadium specific manner[5;6]. According to this, the genetic control of the human CD83 gene provides an interesting subject for studying and understanding immune cell specific regulation mechanisms. Furthermore, gaining an insight in the genetic regulation of the CD83 gene may provide crucial information to reveal the function of this interesting molecule in the immune system. Moreover, a mature DC-specific promoter would be a valuable tool for DC-mediated vaccination strategies to express therapeutic proteins specifically only in mature DCs (mDCs) to elicit e.g. a tumor-specific immune response, as the targeting of immature DCs (iDCs) will induce tolerogenic mechanisms. To this end, combining said regulation mechanism with the efforts of DC based immunotherapy to specifically guide DCs to induce an immune response against a desired target, particularly cancer cells, remains a task of great interest and challenge for the future.

2.1. The biology of dendritic cells

2.1.1. DC subsets

The DC-system consists of distinct subsets. Although all of them function as APCs, the several and often opposing roles ascribed to DCs cannot all be carried out at once by the same cell[7]. Regarding to this, different subsets of DCs have been described recently in the murine as well as in the human system.[8] Originally, murine splenic DCs were separated into $CD8\alpha^+$ $CD11b^-$ and $CD8\alpha^-$ $CD11b^+$ major subsets, each expressing $CD11c^9$. The $CD8\alpha^+$ subpopulation was initially thought to derive exclusively from $CD34^+$ lymphoid precursors, hence they were named lymphoid DCs (L-DCs), whereas the $CD8\alpha^-$ cells were thought to derive from $CD34^+$ myeloid precursors and therefore named myeloid DCs (M-DCs). This turned out not to be fully correct, as it has been shown recently that precursors from both lineages could develop into different subtypes of mature splenic and thymic DCs[10-13]. Murine $CD8\alpha^+$ DCs express CD1d and C-type lectin DEC-205 and have the ability to cross prime $CD8^+$ T cells as well as to produce large amounts of interleukin-12 (IL-12) to induce preferentially T helper 1 (Th_1) responses[14-18]. Murine $CD8\alpha^-$ cells on the other hand lack IL-12 expression, induce preferentially T helper 2 (Th_2) responses and do not cross prime $CD8^+$ T cells[19]. They reside in the marginal zone of the spleen, but migrate into the T cell zone upon microbial stimulation[20;21]. Beyond that, the $CD8\alpha^-$ $CD11b^+$ subset of DCs can be further subdivided into a $CD4^+$ and $CD4^-$ subpopulation with yet unknown functions[9]. Besides the $CD8\alpha^+$ and the $CD8\alpha^-$ DC subsets, a third major subpopulation has been identified, the plasmacytoid DCs (P-DCs), which are characterized by their plasmacytoid appearance, lack of CD123, low CD11c, high B220 and Gr-1 expression. In response to viral or other microbial stimuli they produce high levels of type one interferon (IFN) and upregulate several surface markers like CD80, CD83 and CD86 in both mice and humans[22-24].

However, in the human system the DC subsets are less well understood. Two major subsets of human DCs, which represent only 0.3~1% of all circulating leukocytes[25], were identified based on the expression of $CD11c^{26}$. Both subsets derive from $CD34^+$ precursors and are - like their murine counterparts - defined

Introduction

on the one hand as myeloid DCs, expressing CD11c and on the other hand as plasmacytoid DCs, lacking CD11c expression[27-30]. Further *in vitro* investigation revealed two additional dermal M-DC subtypes[31]: CD1a⁺ CD14⁻ Langerhans cells (LCs) expressing CD1a, Langerin and E-cadherin as well as CD1a⁻ CD14⁺ *interstitial* DCs (intDCs) expressing DC-SIGN, CD11b and factor XIIIa[32;33]. Both subpopulations reside in different layers of the skin (epidermis and dermis respectively), show different migration kinetics and differ in phenotypes and biological functions[34]. LCs preferentially induce cellular immunity as they efficiently promote the differentiation of naive CD4⁺ T cells into activated cells secreting IFN-γ, IL-4, IL-5 and IL-13 and provide help to CD8⁺ T cells[35]. In contrast, intDCs have been described to produce IL-10 in response to CD40L stimulation[36] and to induce preferentially humoral immunity by promoting differentiation of naive B cells into plasma cells by secretion of IL-6 and IL-12[37;38]. In the blood of healthy donors a third myeloid DC subset and two plasmacytoid DC subtypes, CD2⁻ and CD2⁺ P-DCs, have been identified recently: The blood M-DCs are characterized as lineage marker negative (Lin⁻) and HLA-DR⁺ cells with an expression of CD11c[39;40]. To segregate these cells from skin M-DCs, these DCs were termed conventional DCs (cDC). To date, their physiological role is not clear, but they may either represent a reservoir of precursor cells that replenish DC populations or a specialized sentinel against blood borne pathogens. Blood M-DCs can further be subdivided by the expression of CD1a, CD16 or CD34[41;42].

Human plasmacytoid DCs in general secrete large amounts of type one IFN[43;44] as well as other cytokines in response to viral infections. P-DCs have been shown *in vitro* to function as APCs and to induce Th₁ and to a lesser extend also Th₂ responses[45]. Furthermore they activate cytotoxic CD8⁺ T cells[46], but their role in priming T cell responses is not fully understood yet. Two human P-DC subsets, CD2⁺ and CD2⁻, were distinguished so far[47]. CD2⁺ P-DCs express high amounts of lysozyme, CD80 and IL12p40 and efficiently kill target cells in a TRAIL-dependent fashion. Moreover, they are more efficient than CD2⁻ P-DC in inducing proliferation of allogeneic naive CD4⁺ T cells[46;47]. CD2⁻ P-DC on the other hand do not efficiently kill target cells, as they express almost no lysozyme like their CD2⁺ counterparts and show almost no cell clustering with

Introduction

their target cells. Moreover, they express only low levels of CD80 and IL12p40[47].

Recently, human equivalents of murine CD8α[+] DC have been described by the groups of Bachem, Crozat and Poulin[48-50]. These leukocytes express the chemokine receptor XCR1 as a conserved selective marker[48-50]. Furthermore they are characterized by a high expression of the C-type lectins CLEC9a (DNGR-1) and thrombomodulin (BDCA3 or CD141) as well as the integrin CD11c. Like mouse CD8α[+] DCs, human DNGR-1[+] BDCA3[+] DCs express nectin-like molecule 2 (Necl2), CD207, basic leucine zipper transcription factor/ATF-like 3 (BATF3), interferon regulatory factor 8 (IRF-8) and toll-like receptor 3 (TLR3), but not CD11b, IRF-4, TLR7 or (unlike murine CD8α[+] DCs) TLR9. The DNGR-1[+] BDCA3[+] DCs respond to poly I:C and agonists of TLR8 (e.g. G-rich oligonucleotides), but not of TLR7 (e.g. single stranded, endosomal RNA), and produce interleukin (IL)-12 when given innate and T cell–derived signals. Importantly, these cells are better in cross-presentation of soluble or cell-associated antigens to CD8[+] T cells in comparison to CD1c[+] DCs, CD16[+] DCs, and P-DCs from the same donor[48-50]. An overview of the different subtypes of human DCs and their specific molecules is given in figure 2.1.

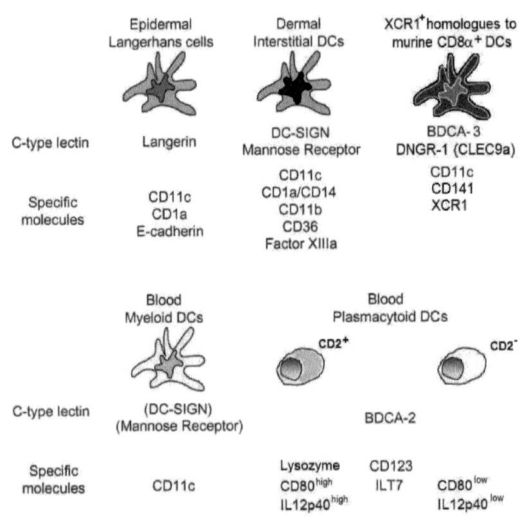

Figure 2.1. Overview on human DC subsets and their main specific molecules (adapted from Ueno et al.[46])

Introduction

2.1.2. DCs as immunological sentinels

Regarding their role as most APCs the task of DCs is to gather information of the surrounding environment by capturing and processing antigens, transferring the information to the various immunological authorities and guiding an apt immune response in terms of quality – either humoral or cellular - and quantity - ranging from tolerance to full blown immunity. Thereby the different subtypes of DCs represent the diversity of the immune response they induce. In general, DCs with an immature phenotype take up antigens until they are activated by a so called "danger signal" leading to a process called maturation. The maturing cell then migrates to the proximate lymphoid organ carrying the gathered information in order to stimulate naive lymphocytes, resulting in an efficient immune response.

2.1.2.1 Antigen capture

Immature myeloid DCs constantly and indiscriminately macropinocytose extracellular fluid and express specific receptors to mediate phagocytosis and endocytosis of pathogens and antigens. DCs express C-type lectin receptors that recognize glycosylated antigens exploiting highly conserved carbohydrate recognition domains (CRDs) in a calcium-dependent manner[51], for example DEC-205, langerin and DC-SIGN, as well as the more common mannose receptor[52-55]. Moreover, C-type lectins also play an important role for the motility of DCs[56]. DCs, like macrophages, also express Fc receptors to mediate the uptake of immune complexes in order to amplify immune responses[57;58]. Furthermore, several hematopoietic cell types, including DCs, express heat-shock protein (hsp) binding receptors that mediate the uptake of hsp-peptide complexes (e.g. with antigenic intracellular peptides)[59]. Finally, the binding and phagocytosis of apoptotic bodies can be mediated by receptors such as CD36 and several $\alpha v \beta 5$ integrins[60]. This vast diversity of surface receptors equip DCs perfectly for the uptake of a broad range of antigens[61].

2.1.2.2 Antigen processing

DCs prime naive T cells via antigenic peptides presented either on major histocompatibility complex (MHC) class I (every nucleated cell) or MHC class II molecules (only APCs) after proteolytic cleavage[62]. MHC class I molecules typically present endogenous peptides, like cytosolic or endoparasite derived proteins, which are loaded onto the molecule in the endoplasmatic reticulum. MHC class II molecules on the other hand typically present extracellular antigens, hence non-cytosolic proteins that are endocytosed, digested in lysosomes, and bound by the MHC class II molecule prior to the molecule's migration to the plasma membrane[63;64].

For class II presentation, DCs cleave the antigen after uptake into 12-25 amino acid (aa) long peptides that are loaded onto the MHC class II molecules in endosomal compartments[65]. At first, the newly synthesized MHC class II molecules, consisting of an α- and a β-chain, are associated with a 31-34 kDa invariant chain (Ii) when expressed on the cell surface and by this blocked for antigen loading[66]. These complexes are then internalized for Ii degradation by cathepsin S and subsequently loaded with peptide[57]. After being recycled to the cell surface, the MHC class II/peptide complexes are degraded in lysosomes. Upon induction of maturation, both the synthesis of new MHC class II molecules and their peptide loading increases with a contemporarily decrease of MHC class II/peptide complex internalization and degradation[67]. This leads to a stabilization of peptide loaded MHC class II molecules on the surface of DCs. Interestingly, another mechanism of MHC class II peptide loading has been observed, where antigens are processed extracellularly by secreted proteases and then loaded efficiently onto empty MHC class II molecules on the surface of iDCs, rising efficiency of antigen uptake by a yet unknown mechanism[68-70].

In the case of MHC class I molecules, that are loaded with peptide in the endoplasmatic reticulum, the peptide presentation is generally restricted to cytosolic, thus endogenously synthesized peptides, which are presented to $CD8^+$ cytotoxic T cells (CTLs)[71-73]. On the one hand this enables CTLs to kill e.g. virus infected cells that present actively synthesized viral protein on their surface, but on the other hand protects cells that have only phagocytozed viral protein from apoptotic bodies and are not infected themselves[71-73]. DCs can

Introduction

present like any other nucleated cell endogenously synthesized protein to CTLs. However, in order to initiate an effective $CD8^+$ CTL response, DCs have to be able to present extracellular antigen on MHC class I molecules to prime naive $CD8^+$ T cells to induce a cellular immune response. This is achieved through a mechanism that is referred to as cross presentation, where APCs take up, process and present extracellular antigens on MHC class I molecules[71;74;75]. DCs are among other APCs particularly efficient for cross presentation[76;74;77].

2.1.2.3. Activation of DCs

For maturation and efficient priming of T and B cells, the activation of DCs is a necessary prerequisite. Several stimuli or danger signals derived from the pathogen itself or provided by elements of either the innate or adaptive immunity cause DCs to respond differently in order to orchestrate the most effective immune response. The most powerful stimulus for DC activation is the engagement of the large array of pathogen sensors called pattern recognition receptors (PRRs) that recognize the so called pathogen-associated molecular patterns (PAMPs)[78]. PAMPs include flagellin and lipoproteins like lipopolysaccharide (LPS) of microbial origin, double stranded RNA of viral origin and beta-glucans of fungal origin, thus covering the most important classes of pathogens[79;80]. Two classes of PRRs have been identified: membrane bound and cytosolic PRRs. The membrane bound PRRs include the Toll-like receptor family, C-type lectins (e.g. the Mannose receptor) and the scavenger receptors (SR), like SR-A or CD68. The cytosolic PRRs are mainly composed of NOD-like receptors (NLRs) like NOD1/2 and RNA helicases, like RIG-I and MDA5 (melanoma differentiation-associated gene 5), which recognize intracellular parasites and viruses[81;82]. Taken together, different activation signals lead to the induction of different cellular responses and either to innate or adaptive immunity.

2.1.2.3.1. Innate immunity activation signals

As a first line of defense the innate immune system provides an initial barrier by either (i) physical means, like the skin or the mucous membranes, (ii) chemical

25

Introduction

means like the acid layer of the skin or the complement system and (iii) the phagocytes of the innate immune system, like macrophages and granulocytes that directly recognize invading pathogens by their PAMPs[82]. Next, the adaptive immune system is activated to efficiently clear the infection. In this regard, DCs provide the link between innate and adaptive immunity by being able to either directly recognize pathogens via evolutionary conserved recognition patterns or indirectly by other players of the innate immune system, such as the complement cascade, macrophages and granulocytes. In either case DCs transmit the signal to the elements of the adaptive immunity. Thereby, one important feature of DCs is the expression of Toll-like receptors (TLR)[83;84]. To date, 11 human TLRs and 13 mouse TLRs have been identified that recognize a variety of PAMPs. Different DC subsets express different sets of TLRs: human myeloid DCs and mouse cDCs express TLR2, TLR3, TLR4, while TLR7, TLR8 and TLR9 are expressed on human and mouse plasmacytoid DCs[78;85]. By recognizing the microbial components, TLRs trigger a cascade of events leading to the activation of specific gene transcription including proinflammatory cytokines and interferons that orchestrate innate immunity as well as chemokines and co-stimulatory molecules that promote T cell activation[81]. All TLRs contain the Toll-interleukin-1 receptor domain (TIR), which recruits one or more TIR-containing adaptor proteins to transmit signals downstream: the myeloid differentiation primary response protein 88 (MyD88), the TIR domain-containing adaptor inducing IFN-β (TRIF, also known as TICAM1), the MAL protein (also called TIR domain-containing adaptor protein [TIRAP]) and the TRIF-related adaptor molecule (TRAM, also called TICAM2)[86;87]. The MyD88 adaptor is recruited by all TLRs except TLR3, which signals only through TRIF. After ligation, TLRs transmit a signal either via the myD88-dependent pathway (TLRs 1, 2, 4-9) or via the MyD88-independent pathway (TLR3 and in some cases TLR4). The MyD88-dependent pathway acts through IL-1 receptor kinase associated kinase 1 and 4 (IRAK1 and 4) and receptor associated factor 6 (TRAF6) leading to the activation of NFκB, IRF-7 or the mitogen activated kinase (MAPK) pathway. Alternatively, the MyD88-independent pathway acts through TRIFs and TRAFs, leading to the activation of IRF-3 or IRF-7[79-81;83;87-89]. Finally, the activation of the NFκB-, IRF- and MAPK-pathways results in the synthesis of proinflammatory cytokines and type I IFN, which can then act in an

Introduction

autocrine and paracrine manner to further activate DCs and shape the subsequent adaptive immune response. In either way, the ligation of TLRs leads to the activation and maturation of DCs followed by the induction of a primary immune response, including the upregulation of costimulatory molecules and the expression of proinflammatory cytokines like IL-12[90;91]. For an overview of the human TLRs, their ligands and signaling see figure 2.2.

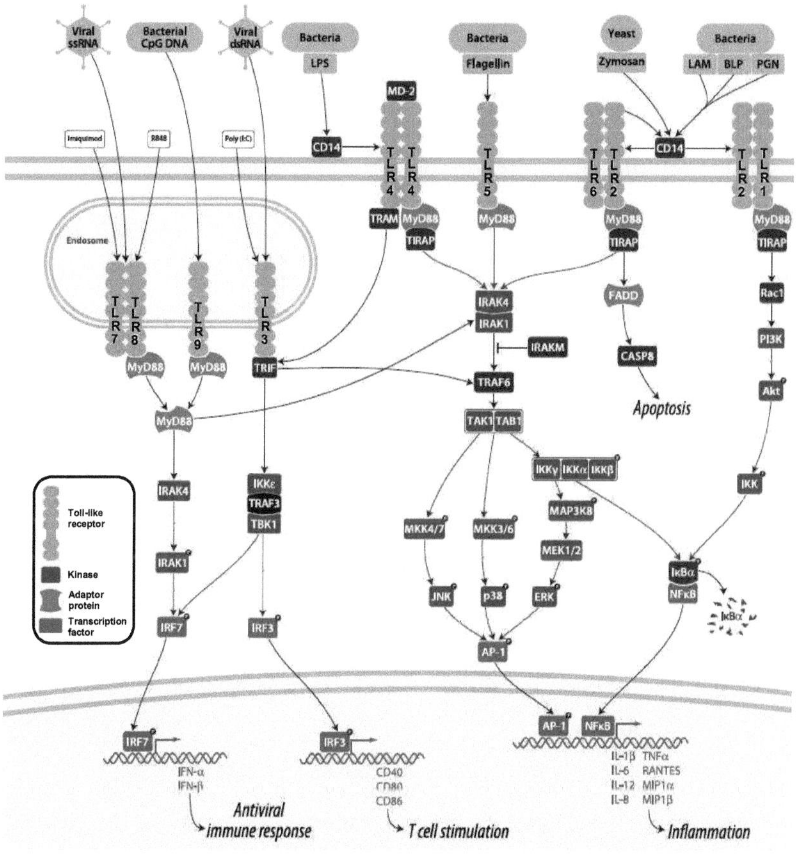

Figure 2.2. Schematic outline of the various TLRs, their ligands and signaling cascades (taken from www.invitrogen.com). After ligation TLRs transmit a signal via the MyD88-dependent pathway, through IL-1 receptor kinase associated kinase (IRAK) and receptor associated factor 6 (TRAF6) leading to the activation of NFκB, IRF-7 or the mitogen activated kinase (MAPK) pathway. Alternatively, mediated through TLR3 and in some cases TLR4, the MyD88-independent pathway acts through TRIF and TRAF activating IRF-3 or IRF-7. TLRs 1, 2, 4, 5 and 6 are expressed on the cell surface; TLRs 3, 7, 8 and 9 are located in endosomes.

Introduction

2.1.2.3.2. Adaptive immunity activation signals

DCs are activated by the adaptive immune system by two different mechanisms: CD40-CD40 ligand (CD40L) interaction and Fc (fragment crystallizable region) receptor engagement.

CD40 is a member of the TNF receptor superfamily and is expressed on all APCs. Its ligand CD40L is mainly expressed on activated $CD4^+$ T cells. Engagement of CD40 on the surface of DCs leads not only to DC maturation, but also to an activation process called licensing[92;93]. Moreover, CD40 engagement empowers DCs to activate $CD8^+$ CTLs, thereby providing an important step for an effective cellular immune response[92;94;95].

Thereby the link between humoral adaptive immunity and DCs is provided by Fc receptors. DCs express several Fc receptors such as FcγR and FcεR that bind the Fc domain of the immune globulins G and E (IgG and IgE), respectively. In terms of functionality Fc receptors are considered as antigen receptors, as they lead to the uptake of antigen-antibody complexes resulting in fully activated mDCs[7;58;96].

2.1.2.4. Maturation of DCs

In order to efficiently orchestrate an immune response, DCs have to undergo maturation, as immature or semi-mature DCs induce immunological tolerance by either driving the development of regulatory T cells (T_{reg}) or leading to T cell anergy[97]. Maturation causes changes in phenotype and functionality of DCs, resulting in the ability to prime naive T and B cells, which cannot be attributed to one single event or molecule, but merely to coordinated quantitative effects and regulations. These phenotypical changes include morphological changes (e.g. the acquisition of dendrites), the loss of phagocytotic receptors, upregulation of MHC class I and II molecules, changes in cytokine (-receptor) as well as adhesion molecule expression profiles. These coordinated events ensure an efficient contact with the target cells and to initiate the appropriate immune responses. Furthermore, the upregulation of several costimulatory molecules such as CD40, CD80, CD83 and CD86 allows precise cell to cell communication. Finally the changes in the adhesion molecule pattern ensure

Introduction

prolonged cell to cell contact as well as increased motility to reach the site of immune cell priming[61;98]. Taken together, these features make DCs to the most potent APCs of the immune system.

2.1.2.5. Migration of DCs

After activation and the initiation of maturation DCs migrate from the site of activation to the proximate lymph node. Here DCs (i) attract B and T cells by the released chemokines, (ii) maintain their viability by cytokine release, (iii) induce proliferation and differentiation and (iv) finally mediate an effective immune response[1;99]. Upon encountering other lymphocytes in the lymph node, DCs themselves receive additional activation signals that enable them to mount a $CD8^+$ CTL response. Migration is a tightly regulated part of the maturation process that includes several cytokines, such as GM-CSF, TNF-α, IL-1 and MIP-1α/β), chemokines like CCL19 and CCL21 as well as non-chemokine agonists like PGE_2, lipid mediators and membrane proteins like integrins[100]. Receiving the activating stimulus (through TLR-triggering or CD40 ligation), DCs upregulate chemokine receptor 7 (CCR7)[101] that guides DCs along a chemokine gradient of its ligands (mainly chemokine ligand 19 and 21 [CCL 19/21]) to the lymphatic vessels and the T cell areas of the secondary lymphoid organs. However, CCR7 expression alone is not sufficient for DC migration[102]. A variety of other triggers, including lipid mediators, like sphingolipids, eicosanoid derivates and particularly prostaglandin E_2 (PGE_2) are necessary for an efficient migration. It is noteworthy, that iDC are attracted by different chemokines than maturing DCs. Immature DCs are attracted by MIP-1α/β, MIP-5 and macrophage chemoattractant protein 4 and 5, RANTES and TECK. In contrast, mDCs have lost their responsiveness to these chemokines, but have acquired susceptibility to CCL19 and CCL21[103].

2.1.2.6. Control of different T cell responses by DCs

DCs orchestrate the appropriate immune response through a complex crosstalk between the various elements of the immune system mediated by a wide array of cytokines. In this context, DCs can induce various types of $CD4^+$ T helper cells (Th) responses by the presentation of antigenic peptides on MHC class II molecules: (i) Th_1 cells to induce a cellular immune response[104;105], (ii) Th_2 cells to induce a humoral immune response[106], (iii) regulatory T cells (T_{regs}) to induce immunological tolerance[107], (iv) Th_{17} cells to promote among other functions the recruitment of neutrophils[108] and (v) Th_{22} cells[109] secreting high amounts of IL-22 to induce the innate epithelial immune response. Moreover, DCs also (cross)prime $CD8^+$ CTLs via MHC class I to promote cytotoxic cell killing[110].

The Th_1 or cellular immune response is mainly directed against intracellular pathogens. Priming of $CD4^+$ Th_1 cells occurs via IL-12 produced by DCs. $CD4^+$ Th_1 themselves then produce IL-12, IL-2, INF-γ and TNF-β to stimulate mainly macrophages and NK cells and to provide cytokine mediated help to CTLs[90;91].

Th_2 responses are directed against extracellular pathogens and thus mediate the B cell-driven antibody-dependent immune reaction. The mechanism of Th_2 cell induction by DCs is still under discussion, but it has been described recently that the lack of IL-12 and the OX40 receptor plays an important role. Th_2 cells produce high amounts of IL-4, IL-5 and IL-13. Thereby, IL-4 promotes the Th_2 phenotype and leads to the proliferation and subsequent differentiation of B cells into antibody secreting plasma cells. IL-5 and IL-13 have been described to attract and activate eosinophiles as well as basophiles followed by inflammation[14;106]. The Th_1 and Th_2 reaction control each other reciprocally in such a way that the Th_1 cytokine IL-12 suppresses Th_2 responses and the Th_2 cytokine IL-4 inhibits the Th_1 response[106;111]. Importantly, $CD4^+$ Th cells also support the priming of $CD8^+$ CTLs by CD40 ligation, a mechanism called licensing, closing the circle of complex DC-T cell communication[110;112].

Third, DCs can induce immunological tolerance and/or restrain an ongoing immune response by inducing T_{regs} inhibiting activation of $CD4^+$ and $CD8^+$ T cells. T_{regs} are divided into several subsets: $CD4^+$ $CD25^+$ T_{regs}, T_R1 and Th_3 cells, which mediate the restriction of the immune response by cell-cell contact mechanisms and/or the secretion of mainly IL-10, *transforming growth factor beta* (TGF-β) and IL-35. Moreover, it has been reported that the transcription factor FOXP3, as well as Indoleamine 2, 3 Dioxygenase (IDO) play an important role in the development of these cells[113-115].

Another kind of T cell type are Th_{17} and Th_{22} cells, which produce high amounts of IL-17 and IL-22, respectively. Both cells have been demonstrated to promote

Introduction

tissue inflammation and to play a key role in autoimmune diseases like psoriasis[116;117]. Th$_{17}$ cell differentiation is mainly promoted by IL-6 secretion by DCs and the expression of the transcription factor RORγt[118]. Th$_{22}$ cells are far less understood, but are under intense investigation[116;119]. Taken together, the direction the immune response takes is defined by the type of T cells primed by DCs, which in turn is dependent of the pathogen itself. However, processes leading to the specific priming of the appropriate T cell subtypes are not fully understood to date, but it is clear that the DC holds the central role in determining the direction of the immune response.

2.1.3. The surface molecule CD83

Amongst many other surface molecules like CD80 and CD86 that are upregulated during DC maturation, one membrane bound glycoprotein and member of the IgG superfamily is of outstanding importance: the molecule CD83. CD83 is to date the best known maturation surface marker for human DCs as it is not expressed on the surface of iDCs, but rapidly upregulated upon induction of maturation[120]. However, CD83 is not exclusively expressed by DCs. Activated monocytes and macrophages also rapidly induce CD83 surface expression. In strong contrast to DCs, surface expression on those cell types is not long lasting and after 8–24 hours CD83 has completely disappeared from their surface[121]. Moreover, also some subsets of T cells[122], B cells[123], granulocyte precursor cells[124], myeolocytes[125], neutrophils[126], as well as thymus epithelial cells[127] express CD83 during their life cycle. Furthermore, CD83 expression was also reported on Hodgkin cells[128] and Epstein - Barr virus transformed lymphoblastoid cell lines[129]. In mice the expression pattern of CD83 is similar to humans. Histological staining, FACS analysis and the CD83 reporter mouse revealed CD83 expression on the surface of murine mDCs as well as on activated B and T cells[130;131;132]. Surprisingly, CD83 mRNA processing and protein expression differ from that of most other molecules: In strong contrast to the vast majority of cellular mRNAs, the CD83 mRNA is not exported from the nucleus to the cytoplasm via the commonly used TAP-pathway, but by the CRM1-mediated nuclear export[133-137]. So far two isoforms have been reported: In addition to a membrane bound form of CD83 (mCD83),

Introduction

a soluble form (sCD83) has been described in the sera of healthy donors[138]. Patients suffering from hematological malignancies or rheumatoid arthritis show elevated levels of sCD83[139;140]. The origin of sCD83 is still under debate. Dudziak and colleagues have identified four different transcript isoforms of CD83 in unstimulated peripheral blood mononuclear cells (PBMC) with the longest transcript coding for mCD83. The shorter isoforms may code for soluble splice variants of full length CD83[141]. In mice only a single isoform of mRNA has been reported to date[142]. An alternative hypothesis was formed by Hock and colleagues. They proposed a proteolytic cleavage of mCD83 from the cell surface, thus generating sCD83[143]

2.1.3.1. Structure of CD83 gene and protein

The human CD83 gene is located on chromosome 6p23 and composed of five exons and four introns spanning over a total length of ~19 kb[144]. Likewise, the murine CD83 gene spans over a total length of ~19kb, consisting of five exons and four introns and is located on chromosome 13 band A 5. CD83 is expressed in most, if not all, vertebrate species. It has been reported to be expressed by humans, chimpanzee, horse, swine, cattle, panda, dog, rat, mouse, frog, elasmobranch and teleost fish, sharing the highest amino acids (aa) sequence homology with chimpanzee (*pan troglodytes*, 99%), horse, (*equus caballus*, 76%), cattle (*bos taurus*, 74%), swine (*sus scrofa*, 72 %) and mouse (*mus musculus*, 65%) and to a lesser extend with fish (*oncorhynchus mykiss* and *ginglymostoma cirratum*, 28% and 32% respectively)[145;146].

The human CD83 protein is 205 aa long with a molecular weight of ~45 kDa and belongs to the immune globulin superfamily exhibiting a single extracellular V-type Ig like domain[5]. Molecular analyses revealed that aa 1-19 encode the signal peptide, whereas aa 20-144 form the extracellular domain. Four cysteines at amino acid positions 27, 35, 100 and 107 have been shown to form intramolecular disulfide bonds (aa 27-100; aa 35-107) leading to the V-type Ig-like domain. A fifth cysteine at position 129 is involved in the formation of an intermolecular covalent disulfide bond leading to the dimerization of the extracellular protein domains[147]. Amino acids 145-166 form the helical transmembrane domain and aa 167-205 compose the intracellular domain.

Furthermore CD83 is post-translationally glycosylated at aa positions 79, 86, 96 and 117[148].

The murine CD83 is 175 aa long and shares ~63% of its amino acid sequence with the human protein thereby featuring a similar protein domain architecture[131;142]. One major difference between the structures of human and murine CD83 is that aa 65-75 from the human CD83 are missing in the murine CD83[145]. Iterative sequence database searches using PSI-BLAST suggest that both molecules represent divergent members of the immunoglobulin (Ig) family[145]. However, the positions of all five cysteine residues are well conserved between mice and humans suggesting the same inter- and intramolecular disulfide bonds[145].

2.1.3.2. Functions of membrane bound and soluble forms of CD83

Membrane bound CD83 has been suggested to act as an essential enhancer/costimulatory molecule during T cell activation, as electroporation with small interfering RNA (siRNA) for CD83 knock down resulted in a clearly reduced T cell activation by DCs[149]. These data are in agreement with another siRNA study by the group of Aerts-Toegaert, who clearly showed that the allogeneic T cell stimulatory capacity of mDC decreases after CD83 siRNA treatment[150]. In this context, Kruse and colleagues showed that the nucleocytoplasmic translocation of the CD83 mRNA in human monocyte derived DCs (moDC) can be inhibited by interfering with the hypusine modification of the eukaryotic initiation factor 5A (eIF-5A), which is part of a of a particular RNA nuclear export pathway[151]. Interestingly, this significantly inhibited DC-mediated T cell activation. Furthermore, analyses of CD83$^{-/-}$ mice pointed out that the CD83 signal is required both from thymic endothelial cells on a transitional CD4$^+$/CD8^{low} thymocyte population as well as from DCs during the development of thymocytes[127].

Soluble CD83 on the other hand has been shown to exert immunosuppressive functions. Incubation of human sCD83 with DCs interfered with cocktail induced maturation leading to an incomplete matured phenotype[152]. Furthermore, sCD83 showed a severe effect on the cellular cytoskeleton of maturing DCs i.e. the cells rounded up, failed to develop proper veils and were severely inhibited

Introduction

in their ability to form clusters with T cells[153]. Moreover, sCD83 inhibits DC mediated T cell stimulation in a concentration dependent manner *in vitro*[154]. Furthermore, *in vivo* human sCD83 has been demonstrated to protect mice from the experimental autoimmune encephalomyelitis (EAE), a CD4⁺ T cell mediated model for the early inflammatory stage of human multiple sclerosis[155]. Moreover, several transplantation experiments have shown the immunosuppressive properties of human soluble CD83: (i) In combination with the immunosuppressive drug Rapamycine and anti-CD45RB antibody sCD83 induced transplant tolerance in a C3H-to-C57BL/6 mouse cardiac transplantation model[156]. (ii) Studies showed that a fusion protein consisting of the extracellular part of murine CD83 and a human IgG1α Fc tail delayed acute cellular rejection of MHC-mismatched skin allografts in mice and inhibits antigen-specific T cell proliferation and IL-2 secretion in spleen cell cultures from DO11.10 T cell receptor transgenic mice[157]. (iii) Rats treated with sCD83 exhibited the prevention of chronic renal allograft rejection[158].The treated animals showed a marked decrease in IgM and IgG deposition in the graft and a reduced infiltration of T cells and monocytes into the graft tissue. (iv) It has been shown that sCD83 induces tolerance for kidney allografts in mice[159]. Splenic DCs of treated mice exhibited significantly decreased levels of surface MHC II, CD40, CD80, and intracellular interleukin-12, as well as reduced allogeneic stimulatory capacity prolonging kidney allograft tolerance for more than 100 days. Moreover, chronic lymphocytic leukemia (CLL) patients with elevated plasma sCD83 levels showed significantly shorter treatment free survival[160]. Furthermore, Scholler and colleagues could show that injection of fusion proteins obtained by fusing the extracellular domain of human CD83 to human or murine Ig tails (CD83-Ig) inhibited the development of CTLs in mice *in vivo* and human cells *in vitro* [154]. The same group further demonstrated that CD83-Ig is costimulatory when co-immobilized with anti-CD3 and that cells from a mouse melanoma transfected with CD83 induce a tumor rejection response against wild-type tumor cells. Surprisingly, a murine equivalent to the human sCD83 has not been described yet. Interestingly, a CD83 ligand has not yet been characterized and its existence is still under controversial discussion[132;152;161]. However, the presence of a yet unknown CD83 ligand

Introduction

(CD83L) on murine B220⁺ B cells as well as on human T cells has been suggested by Cramer and Hirano[132;162].

2.1.3.3. CD83 as a target for viral immune escape mechanisms

The outstanding importance of CD83 for the function of DCs and the immune response in general is further highlighted by the finding that CD83 is a direct target of viral immune escape mechanisms of human Cytomegalovirus (HCMV)[163] and Herpes simplex virus type-1 (HSV-1)[164]. Both viruses inhibit the biological function of CD83 to avoid antiviral immune responses. On the one hand, HCMV is able to infect both iDCs and mDCs and thereby inhibits the stimulation of T-cell proliferation[163;165;166]. Interestingly, the infection with HCMV led to the shedding of a soluble form of CD83 by proteolytic cleavage and a subsequent block in T-cell stimulation[154].

On the other hand, after the infection of mDCs with HSV-1, a strong downregulation of the cell surface expression of CD83 was observed, which did not occur due to shedding of CD83 from surface, but due to a very fast and efficient degradation of the CD83 molecule inside the infected DCs[164]. Already 10 hours after infection CD83 was almost completely removed from the cell surface, whereas other co-stimulatory molecules such as CD80 and CD86 were not influenced and were still expressed at high levels on the cell surface. Nevertheless, the CD83 down-modulation correlated with a reduced T-cell stimulatory capacity[164].

In summary, these findings have not only provided insights into the function of the CD83 molecule, but also prospects for the establishment of new therapeutic applications in the treatment of autoimmune diseases, infectious diseases, T-cell based allergic reactions, transplant rejection and immune therapy for malignant tumors.

2.1.3.4. The CD83 promoter

In 2002 Susanne Berchtold and colleagues published a human CD83 minimal promoter (MP -261) sequence[167]. They cloned a 3037 bp long fragment upstream of the transcription start codon and narrowed the core promoter

Introduction

sequence down to 261 bp, using deletion mutagenesis. Bioinformatical analyses revealed four specificity protein 1 (SP1) and one NFκB binding element, which could be verified by electrophoretic mobility shift assays (EMSA). These findings are in concordance with those from other groups: McKinsey and colleagues described that NFκB regulates inducible CD83 gene expression in activated T lymphocytes[168] and Dudziak and coworkers published that Epstein Barr virus' latent membrane protein 1 (LMP1) induces CD83 expression in B cells via the NFκB-signaling pathway[129]. Interestingly, Berchtold and coworkers found similar activity of the MP -261 in several other cell types than DCs. The promoter displayed a relatively high activity in mature human moDCs, but also in the human monocyte cell line U937 as well as in human leukemia Jurkat T cells (both expressing CD83) and in the murine DC-like cell line DC2.4.

The exact promoter sequence of the murine CD83 has not been identified yet, but consistent with the human CD83 promoter region, the murine 5' region of the CD83 gene lacks a clear TATA box sequence. However, no conservation of specific transcription factor binding sequences between mice and human were found yet[142].

In summary, the human 261 bp CD83 minimal promoter showed neither cell type nor maturation specific activity for DCs. This lead us to the conclusion that additional regulatory elements must be involved in the cell type- and stadium-specific regulation of the human CD83 gene expression, which have not been identified yet. Thus, the aim of the thesis was the functional characterization of the human DC-specific CD83 promoter.

2.1.4. Adenoviruses and gene therapy

For the transduction of DCs *in vitro*, adenoviral vectors are of great advantage, as they (i) have the ability to grow recombinant viruses to high titers, (ii) the virus displays a high capacity for transgene insertion and an efficient transduction of both resting and dividing cells and (iii) adenoviruses usually do not incorporate their viral DNA into the host genome, but persist as episomes[169-171].

Introduction

Adenoviruses are non-enveloped virions of 70-90 nm in diameter. Their capsids consist of three major proteins: The hexon, the fiber and the penton (Fig. 2.3.). The 240 hexons account for the majority of the structural components of the capsid, the 12 pentons form the 12 vertices of the capsid from which the fibers protrude. These fibers tether the viral capsid to the surface of the host cell by interacting with a cellular receptor. The adenoviral core contains the 36 kb long viral genome consisting of linear double-stranded DNA as well as several proteins including pV and the histone-like pVII that attach the viral genome to the capsid[172].

The entry of the adenovirus into the host cell is achieved by a two-step binding mechanism. The first step is mediated through the binding of the fiber knob to several cellular receptors, depending on the adenovirus serotype and the host cell. Initial attachment either occurs via the coxsackie adenovirus receptor (CAR) or, in the case of DCs, via DC-SIGN, CD80 and CD86[173]. In the second step, after the initial binding, the penton protein interacts with $\alpha_v\beta3$ or $\alpha_v\beta5$ integrins[174;175] on the host cell surface, resulting in receptor-mediated endocytosis of the virus via clathrin-coated pits[176;177]. The virus then escapes from the endosome and enters via the nuclear pore complex. Next, the viral DNA is transported to the nucleus, where gene expression, viral replication and assembly can occur[178;179].

Adenoviruses used for gene transfer often have a modified tropism[180] and are defective in their E1 and E3 genes to ablate their replication capacity and to allow to accommodate up to 8.2 kb foreign DNA into the vector[181]. These so called first generation adenoviral vectors could only be grown in a specifically modified 293 cell line that provided the E1 protein *in trans*[182]. Although being highly promising as gene delivery vehicles, several problems arose with the first generation adenoviral vectors: First, recombination of the virus genome and the E1 region of the complementing cell line gave rise to a replication competent virus generation[183]. Second, the first generation adenoviruses have been shown to stimulate cellular immune responses by their low level gene expression, resulting in the destruction of the transduced cells that were expressing the therapeutic transgenes[184;185]. To prevent both the generation of replication competent adenoviruses and the immune response, so called second generation adenoviral vectors have been created. These vectors additionally lack the E2 and E4 genes, thereby providing a larger capacity for transgene insertion and a reduced viral gene expression[186]. Clinical studies in humans with adenoviruses have already been conducted[187;188].

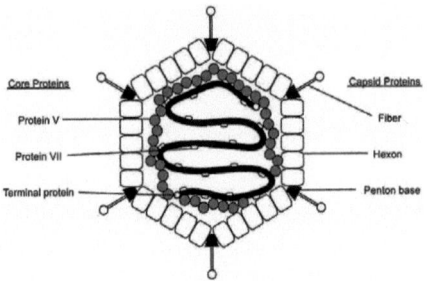

Figure 2.3. Schematic outline of the adenovirus particle (adapted from Mc Connell 2004[172]).

2.2. Genetic regulation and expression mechanisms

In order to characterize the DC-specific CD83 promoter, an understanding of the basic mechanisms of genetic regulation and expression mechanisms is of fundamental importance. Although every cell has the same set of genes, expression between different cell types differs significantly. Cells only express specific genes at a defined time point or as reaction to a certain stimulus to fulfill their particular task. To achieve this tempo-spatially separated gene expression, several levels of gene regulation exist: Every step of gene expression can be modulated, ranging from transcription to post-translational modification of a protein. Additionally, chromatin accessibility, mRNA transport and mRNA stability are subjected to regulatory mechanisms. Moreover, the expression of specific enzymes (e.g. methyltransferases, acetyltransferases and transcription factors) are regulated themselves.

Introduction

2.2.1. Structural modifications of DNA

Epigenetics study mainly inherited changes in gene expression caused by mechanisms other than changes in the underlying DNA sequence. On the one hand the DNA can be directly modified in a chemical way, most notably through methylation. On the other hand the chromatin conformation, structure and thus the accessibility and therefore transcription can be modified by the so called histone code. Both modifications are linked and hence essential for differential gene expression.

2.2.2. DNA Methylation

DNA methylation is essential for normal development of any eukaryotic organism[189] and is associated with a number of key processes including genomic imprinting[190], X-chromosome inactivation[191], suppression of repetitive elements[192] and carcinogenesis[193;194]. The process of DNA methylation is characterized by the addition of a methyl groups to the DNA (Fig. 2.4.A). In this regard, two principles are described to date: First, the addition of a methyl group to the fifth carbon of the cytosine pyrimidine ring (5-meC)[195], leading to the specific silencing of gene expression in eukaryotic cells. Second, methylation at the sixth nitrogen of the adenine purine ring (6-meA) has been mainly described for bacteria in the context of enzymatic restriction of DNA. Its role in eukaryotic gene regulation is still under intense investigation[196]. In adult somatic tissues, DNA methylation typically occurs at a cytosine, immediately followed by a guanine and connected by a phosphodiester bond dinucleotide (CpG)[197]. In contrast, non-CpG methylation (like CpA or CpT) is prevalent in embryonic stem cells[198-200]. DNA methylation can be inherited through cell division, but is typically erased during zygote formation and re-established through successive cell divisions during development. Consequently, it is a crucial part of normal organismal development and cellular differentiation in higher organisms. Furthermore, methylation also forms the basis of chromatin structure[201]. DNA methylation has not only been shown to influence tissue-, lineage- and cell type-specific gene expression [202;203], but also to be important during the life cycle of a single cell[204]. Between 60% and 90% of all CpGs are methylated in

Introduction

mammals[195;205] and often clustered in so called CpG islands present in the 5' regulatory regions of many genes. On the other hand promoter distal CpG sequences gain growing attention, as they seem to play an important role in cell type- and lineage-specific methylation. CpG dinucleotides are underrepresented in the human genome (they occur with only 21% of the expected frequency) [206], as methylated cytosine residues spontaneously deaminate to form thymine residues, hence CpG dinucleotides steadily mutate to TpG dinucleotides. Furthermore, spontaneous deamination of unmethylated cytosine residues gives rise to uracil residues that are quickly recognized by the cell repair machinery and finally eliminated. In many disease processes such as cancer, gene promoter CpG islands acquire abnormal hypermethylation, which results in transcriptional silencing (e.g. of tumor suppressor genes) and can be inherited by daughter cells, thereby representing a target for epigenetic therapy[207;208]. Also hypomethylation has been reported in connection with cancer, but it arises generally earlier in the development and is also linked to chromosomal instability as well as the loss of imprinting[207;209]. Taken together, alterations of DNA methylation have been recognized as an important component of cancer development.

DNA methylation can affect the transcription of genes in two ways. First, the methylation of the DNA itself can physically impede the binding of transcriptional proteins to the gene. Second, methylated DNA can bind methyl-CpG-binding domain proteins (MBDs). MBDs recruit additional proteins to the locus, such as histone deacetylases and other chromatin remodeling proteins that modify histones, thereby remodeling the chromatin[210] (Fig. 2.4. B).

In mammalian cells DNA methylation is carried out by two general types of enzymatic activities: maintenance methylation and *de novo* methylation[211]. Maintenance methylation activity is necessary to preserve DNA methylation after every cellular DNA replication cycle and is carried out by DNA methyltransferases (DNMT), as the semiconservative replication would produce unmethylated daughter strands, thus leading to passive demethylation over time[212;213]. DNMT1 is the most abundant methyltransferase and is responsible for maintenance of DNA methylation by copying the methylation patterns to the daughter strands during DNA replication. *De novo* methylation on the other

Introduction

hand is carried out by DNMT3a and DNMT3b that set up the genome wide methylation patterns in early embryonic development [189].

Taken together, CpG methylation constitutes an important epigenetic mechanism to silence gene expression both in a hereditary as well as highly dynamic manner.

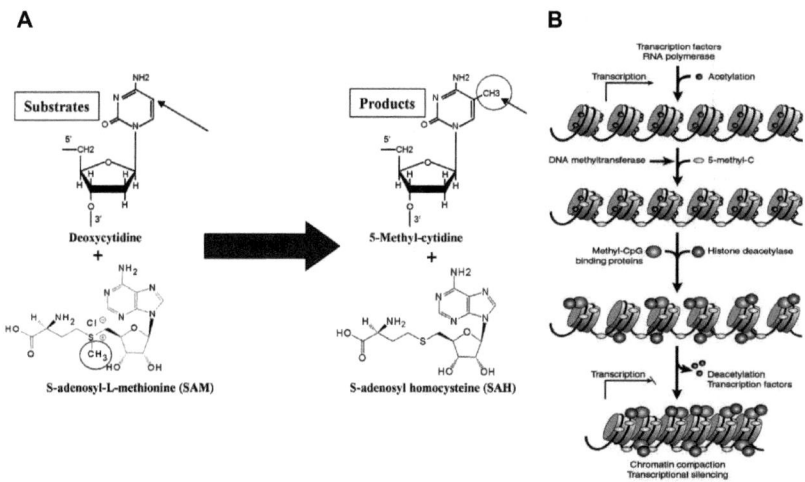

Figure 2.4. (A) Conversion of Deoxycytidine to 5-Methyl-cytidine (adapted from http://www.med.ufl.edu/biochem/keithr/ research.html). DNMT, DNA methyltransferase; SAM, S-adenosyl-L-methionine; SAH, S-adenosyl homocysteine. (B) CpG methylation and histone modification cooperate in gene silencing (adapted from http://www.med.ufl.edu/ biochem/keithr/research.html). 5-methyl-C, 5-methyl-cytidine.

2.2.3. Histone modification and chromatin remodeling

The histone code is hypothesized to be a code consisting of covalent histone tail modifications. Together with other modifications such as DNA methylation it is part of the epigenetic code[214]. The main role of histones is the association with DNA to form nucleosomes, which themselves can bundle to form chromatin fibers[215;216]. Histones are globular proteins with a flexible N-terminus (tail) that protrudes from the nucleosome. Roughly two helical DNA turns wrap around an octamer of core histone proteins formed by four histone partners: a H3-H4

Introduction

tetramer and two H2A-H2B dimers. The fifth histone H1 resides at the entry and exit of the nucleosome stabilizing the chromatin structure[217]. The tail modifications play an important role in the chromatin structure, while the chromatin structure plays an important role in regulation of gene expression[218]. It is hypothesized that chromatin-DNA interactions are guided by combinations of histone modifications[214]. While it is accepted that modifications of histone tails alter the chromatin structure, a complete understanding of the precise mechanisms by which these alterations to histone tails influence DNA-histone interactions remains unclear. So far five histone tail modifications have been found: methylation, acetylation, ADP-ribosylation, ubiquitination and phosphorylation[219]. The histone code shows plasticity in terms of various possibilities for histone modifications, with methylation and acetylation being the most intensively studied[220].

2.2.4. Histone methylation

Histone methylation can occur as mono-, di- and tri-methylation at lysines (K) on the tails of histone 2, 3 and 4[221]. For long thought to be a permanent modification associated with gene silencing, more recent studies imply a much more complex picture. Depending on the fold of methylation and the specific lysine, methylation can either repress or induce gene repression and is indeed reversible[222]. Histone methyltransferases (HMT) catalyze the transfer of one to three methyl-groups from the cofactor S-Adenosyl methionine to lysine (histone-lysine N-methyltransferase) or arginine (histone-arginine N-methyltransferase) residues of histone tails[223]. The more recently discovered histone demethylases catalyze the removal of methyl groups on histone lysine and arginine residues. Two kinds of histone demethylases have been identified, including lysine specific demethylase 1 (LSD1) and Jumonji C (JmjC) domain family proteins [224-226]. The oxidation reaction catalyzed by LSD1 depends on the cofactor FAD and generates an unmodified lysine (H3K4) and a formaldehyde byproduct[225]. The catalytic mechanisms of JmjC domain proteins are all hydroxylation reactions using Fe(II) and *alpha*-Ketoglutaric acid (α-KG) as cofactors[218]. In summary, histone methylation is a multifaceted event that dynamically enables or disables gene transcription.

2.2.5. Histone acetylation

Histones are acetylated and deacetylated on lysine residues in the N-terminal tail and on the surface of the nucleosome core as a part of gene regulation by histone acetyltransferases (HATs) and histone deacetylases (HDACs), respectively [214;218;227]. The source of the acetyl group during histone acetylation is Acetyl-Coenzyme A, whereas during histone deacetylation the acetyl group is transferred back to Coenzyme A[228]. This acetyl group carries a negative charge that neutralizes the positive charge on the histones and thus decreases the interaction with the negatively charged phosphate groups of the DNA. As a consequence, the condensed chromatin is relaxed to a more bulked structure. Thereby, the more relaxed structure of the DNA entails better accessibility for transcription factors as well as more flexibility for conformational changes in DNA structure [229;230]. This relaxation can be reversed by HDAC activity, where the acetyl group is transferred back to Coenzyme A. Relaxed and transcriptionally active DNA is referred to as euchromatin. More condensed (tightly packed) DNA is referred to as heterochromatin[231;232]. Taken together, histone acetylation is one of the most important histone modifications to enable efficient gene transcription. This ubiquitous and frequently occurring modification has been shown to vary between different tissues and cell types and to be maintained by a dynamic balance between the activities of HATs and HDACs.

2.2.6. Regulation of transcription

The most important part of gene regulation is the effective initiation of transcription, which is controlled on several levels: accessibility of the locus, specific transcription factors and regulatory elements, like enhancer or silencers. These events leading to transcription of eukaryotic protein-coding genes culminate in the positioning of RNA polymerase II at the correct initiation site, which is preceded by multiple events: (i) Decondensation of the locus, (ii) nucleosome remodeling, (iii) histone modification, (iv) binding of transcriptional activators and coactivators to the enhancer and promoter as well as (v) the recruitment of the basal transcription machinery to the core promoter. Hence,

Introduction

several DNA regions are in control of the right positioning of the RNA polymerase II combined with the binding of specific transcription factors to these regions, which mediate the positioning by rather complex protein-protein interactions. Apart from this, several repressive DNA elements as well as repressive transcription factors can negatively influence the rate of transcription as an additional level of regulation by recruiting factors like the negative cofactor 2 (NC2) that interfere with the basal transcriptional machinery[233-236]. In summary, the correct positioning of the RNA polymerase II in order to enable an efficient mRNA transcription is subjected to many regulatory mechanisms and elements.

2.2.7. Regulatory DNA elements

Every functional gene consists of a transcriptional region and a regulatory region. The transcriptional region is transcribed into a primary transcript. The regulatory region can be divided into a cis- and a trans-regulatory element. The cis-regulatory elements are the binding sites of transcription factors which can either enhance or repress transcription. The trans-regulatory elements are the DNA sequences, which encode transcription factors[235]. The cis-acting elements can be divided into four types: (i) Promoter: DNA element, where the transcription initiation takes place[237]. (ii) Enhancer: DNA element that, upon binding of transcription factors, enhances transcription[230;238]. (iii) Silencer: DNA element that, upon binding with transcription factors, can repress transcription[239]. (iv) Response element: Recognition site of certain transcription factors. Figure 2.5. depicts a schematic overview of the organization of an eukaryotic gene.

Introduction

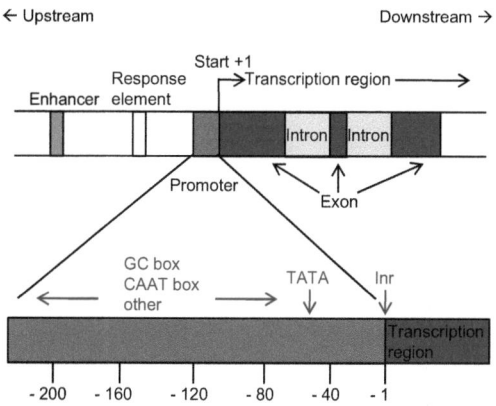

Figure 2.5. Eukaryotic gene organization (adapted from http://www.webbooks.com/MoBio Free/Ch4C.htm). The transcription region consists of exons and introns. The regulatory elements include promoter, response element and enhancer. Downstream refers to the direction of transcription and upstream is opposite to the transcription direction. The numbering of base pairs in the promoter region increases along the direction of transcription, with "+1" assigned for the transcription start. The promoter elements GC box, CAAT box, TATA box and Initiator (Inr) are labeled accordingly.

2.2.7.1. Promoters

The core promoter defines the transcription start site as well as the gene orientation. It includes DNA elements that can extend up to 35 bp upstream and/or downstream of the transcription initiation site[233]. Most core promoter elements appear to interact directly with components of the basal transcription machinery. This basal machinery can be defined as the factors, including the RNA polymerase II itself, that are minimally essential for transcription *in vitro* from an isolated core promoter[237]. The vast majority of studies of the basal machinery have been performed with promoters containing a TATA box as an essential core element, although only ~10% of all promoters contain a TATA box[240]. *In vitro*, TATA-dependent core promoters are associated to the preinitiation complex (PIC) that helps to position the RNA polymerase II over the gene transcription start sites. Furthermore, it denatures the DNA and threads the DNA in the RNA polymerase II active site for transcription[241]. The PIC is composed of six general transcription factors (GTFs) TFIIA, -B, -D, -E, -F, -H

Introduction

and the Mediator complex[237;242]. The GTFs themselves are multi-subunit complexes composed of TATA-box-binding proteins (TBP) and TBP associated factors (TAFs). Transcription then proceeds through a series of steps, including promoter melting, clearance and escape, before a functional RNA polymerase II elongation complex is formed. In addition, core promoters can contain further elements: Initiator Elements (Inr), Downstream Promoter Elements (DPE), Downstream Core Elements (DCE), TFIIB-Recognition Elements (BRE) and Motif Ten Elements (MTE). Except of BRE, all these elements initiate the PIC formation by interacting mainly with TFIID. Approximately 25% of all analyzed promoters contain none of these four elements, suggesting that additional core promoter elements or other types of promoters may occur[233;243;244]. Structural promoter properties that are affected by the underlying nucleotide content and context seem to be important for GTF/DNA interactions. Furthermore, PIC is not dependent on a single assembly point or element, but indeed on the cooperation of several elements[245]. TATA-less core promoters for example interact with TBPs, which are composed of different subunits. This contributes to the regulatory specificity of a gene, as these are influenced by additional cell type specific factors e.g. the response to upstream regulatory elements and associated activators[246].

2.2.7.2. Upstream regulatory elements

As the PIC is only sufficient for initiating low levels of transcription (basal transcription), transcriptional activity has to be stimulated by another class of transcription factors (TF), the so called activators. Activators bind to specific transcription factor binding sites (TFBS) up to 200 bp upstream of the core promoter sequence[247]. These relatively short sequences (6-12 bp) are highly variable and influence the regulatory process by different mechanisms: First, each TF specifically binds to its target sequence. Second, the sequence itself may alter the activity and binding strength of the factor[248]. Third, the factor may bind as a single molecule or as a homo- or heterodimer to the sequence, depending on the expression of interaction partners in a given cell type[247;249;250]. Moreover, activating TFs stimulate and stabilize the PIC assembly by direct interaction with one of the PIC elements or by coactivators[251;252] and promote consecutive steps such as initiation, elongation or reinitiation[241]. Finally, activators may also recruit proteins altering the chromatin structure to enable transcription in general[253;254]. Activators and coactivators are regulated themselves and can act in a stimulating or repressive way ("repressors"). A repressor is a DNA-binding protein that regulates the expression of one or more genes by binding to the response element and blocking the attachment of the RNA polymerase to the promoter, thus preventing transcription of the corresponding genes. Interestingly, also correlations between promoter elements and CpG islands exist[255], further highlighting the complex interactions between transcription factors and DNA structure[240].

2.2.7.3. Enhancers

Enhancers are functionally similar to proximal promoter elements, as they increase the transcriptional activity of promoters and activators. Interestingly, enhancers can act on their specific promoter over great distances of several kilobases and therefore be located either up- or downstream of a gene, within the same or a different gene, or even on a separate chromosome[256;257]. Enhancers regulate gene expression in a tempo-spatial manner and act independently from orientation or distance, although position dependent enhancers have been described[258;259;260]. Tissue-, time point- and stimulus-specificity of promoters often depends on modular enhancers, where a single promoter is induced by different enhancers. Typically enhancers are composed of several clustered TFBSs that work cooperatively. The favored model, how enhancers can act over great distances on promoters, is the looping model. Enhancer and core promoter are brought into close proximity by looping out the intervening DNA[230;238]. Interestingly, studies have also suggested that PIC assembly may be initiated at the distal enhancer and not necessarily at the core promoter[261]. Two models have been proposed of how the TFBSs within the enhancer element cooperate: The "enhanceosome" model and the "billboard" model[262]. The "enhanceosome" model proposes that the TFBSs within the enhancer function as a scaffold to form a unified nucleoprotein complex with a high degree of cooperation between the bound proteins. The function of the enhanceosome is thus more as the sum of its part, but emerges from a network

Introduction

of interactions[260]. In contrast, the "billboard" model suggests that the factors bound to the enhancer do not function as a single unit, but rather interact independently with the basal transcription machinery[263;264]. Taken together, enhancers augment the transcriptional activity of a given gene by several folds, but the underlying mechanisms are still under debate and not fully understood yet.

2.2.7.4. Silencers

Silencers are specific DNA sequences that cause a negative effect on the transcription of a target gene and display most of the properties described for enhancers[265]. Therefore, silencers function independently of orientation and distance from a promoter. It has been also described that silencers are situated as part of a proximal promoter, a distal enhancer or to act as a completely independent module. Furthermore, cooperation and synergistic actions have been reported[266;267]. Silencers can be TFBSs for repressors or exert their repressor function by the recruitment of negative corepressors, which turn a supposed activator into a repressor when interacting with the basal transcription machinery[239]. Moreover, recent findings suggest that silencers can act both in a silencing or in an anti silencing way[268].

2.2.7.5. Insulators

Insulators or boundary elements block genes from being affected by the transcriptional activity of neighboring genes, thus creating discrete islands of expression. They have two main features: First, they can prevent enhancer-promoter communication and second, build up a barrier against condensed chromatin spreading onto active chromatin ("heterochromatin barrier"). Some insulators, like the chicken beta-globin insulator, show these two properties separately[269]. Commonly, insulators are 0.5-3.0 kb long and function in a position and orientation dependent manner. The most well characterized insulator in vertebrates is the chicken β–globin insulator 5'HS4[270]. A homologous element resides in the human β–globin locus[271]. The precise mechanism by which insulators carry out their functions is currently unknown.

Introduction

However, two models exist trying to explain their mode of action[272;273]. In the first model the enhancer blocking activity is described by the inability of insulator-bound activators to carry out their function[230]. The second model associates insulators with the structural organization of chromatin, where it is separated into distinct and independent structural domains by the insulator elements[231]. This model is based on the prerequisite that insulators interact with each other (e.g. two insulators flanking an expression domain or gene) and/or further interact with a nuclear DNA attachment substrate-scaffold, resulting in the formation of loops isolating DNA regions physically from each other.

2.2.8. DNA binding proteins

Efficient gene transcription depends to a large extent on two factors: First, the accessibility of the DNA and second, the correct positioning of the RNA polymerase II. Both events are regulated on multiple levels by the interaction of DNA binding factors with numerous cofactors and their specific binding sites. About 1850 transcription factors (TF) have been described so far, which regulate the expression of 20.000-25.000 genes. Concerning the high diversity of the tempo-spatial expression patterns, it is clear that only the combination of the various DNA binding elements provides the necessary complexity for such a variability in gene regulation[235;274].

2.2.8.1. The role of transcription factors during regulation of transcription

Transcription factors bind directly to their specific DNA sequence and act either as activators or repressors. Numerous coactivators or corepressors mediate between the DNA bound transcription factor and other elements of the cellular regulatory apparatus. The basal transcription machinery, which, together with the Mediator complex is sufficient for basal transcription *in vivo,* is mandatory in most cases. However, many promoters lack the prerequisite to recruit the basal transcription machinery and hence show only weak basal transcription activity. Activators and coactivators provide the prerequisite for an efficient transcription by several mechanisms: First, activating TFs are able to alter the chromatin and DNA structure either directly or indirectly by recruiting DNA modifying enzymes

Introduction

such HATs or demethylases[275;276]. Second, they provide a linker between promoters lacking e.g. a TATA box and GTFs[235]. Third, activators and coactivators enhance the transcriptional efficiency and specificity by several magnitudes through several mechanisms. These include stimulation of PIC formation through a direct interaction with one or more components of the transcriptional machinery or the promotion of initiation, elongation or reinitiation of transcription[251]. Repressors and corepressors on the other hand induce the opposite effect exploiting almost the same mechanism in an inhibitory manner: Altering chromatin structure to a closed conformation, blocking interaction between promoters and GTFs and by suppressing PIC formation[239;277]. During the regulation of immune cell-specific gene expression, especially transcription factors of the specificity protein 1 (SP1) family[278;279], the NFκB-family[280] and the interferon regulatory factor (IRF) family[281] have been shown to play a major role in response to specific stimuli (e.g. PAMPs, cytokines) has shown to regulate the immune cell specific gene expression, thereby determining the outcome of a specific immune response.

2.2.9. The transcription factor SP1

SP1 (specificity protein 1) is an ubiquitously expressed, prototypic C_2H_2-type zinc finger-containing DNA binding protein activating or repressing gene transcription in response to a wide variety of physiologic and pathological stimuli. SP1 was initially characterized as a host factor from HeLa cells binding in a specific manner to the Simian virus 40 (SV40) early promoter[282] and has since been shown to regulate the expression of thousands of genes implicated in the control of a diverse array of cellular processes, such as cell growth[283;284], differentiation[285], apoptosis[283], angiogenesis[286] and immune response[287]. SP1 is a 785-amino-acid, 100- to 110-kDa nuclear transcription factor regulating gene expression by multiple mechanisms. It targets GC-rich motifs (such as 5'-G/T-GGGCGG-G/A-G/A-C/T-3' or '5-G/T-G/A-GGCG-G/T-G/A-G/A-C/T-3') with high affinity[288-290] and has been shown to regulate the expression of TATA-containing and TATA-less promoters by an interplay with other transcription factors, such as Ets-1, c-myc c-Jun, Stat1 and Egr-1, Rel A and/or components of the basal transcriptional machinery[278;291]. Furthermore, SP1 has been linked

Introduction

to chromatin remodeling through interactions with chromatin-modifying factors such as p300[292] and histone deacetylases (HDACs)[293].

In DCs SP1 has been reported to interact with NFκB to regulate the gene expression of CD40[294] and CD83[167]. Tone et al.[294] described that the basal (constitutive) CD40 gene expression is regulated by a TATA-less promoter, with SP1 as a key transcription factor. Likewise, Berchtold et al.[167] showed in luciferase reporter assays that the activity of the TATA-less MP -261 is severely reduced, when any of the four described SP1 sites is deleted from the promoter sequence.

2.2.10. NFκB-family of transcription factors

Over 20 years ago the three proteins, classical NFκB, v-Rel and Dorsal were discovered by the groups of Sen[295], Breitman[296] and Galas[297], respectively. Although showing variable nucleo-cytoplasmic subcellular localization[298-301], the proteins were soon demonstrated to be members of the same family[302-306]. Notably, the biological processes they were involved, namely immunity (NFκB), oncogenesis (v-Rel) and development (Dorsal), evoke much interest in further studying NFκB. Its critical role in different areas of immunology covers Toll-like receptor and antigen receptor (AgR) signaling, lymphoid organogenesis and hematopoiesis [81;307-311] thus, the contribution to the development and survival of cells and tissues that orchestrate the immune response in mammals[280;312;313]. Furthermore, NFκB controls the transcription of cytokines like IL-12[314] and antimicrobial effectors like indoleamine 2,3-dioxygenase (IDO)[315] as well as genes that regulate cellular differentiation, survival and proliferation[280;309;312;316;317], thereby controlling various aspects of innate and adaptive immune responses[300].

This very intensively studied family of transcription factors consists of the "NFκB"-subfamily protein NFκB 1 (p50) and NFκB 2 (p52) as well as the "Rel"-subfamily proteins RelA (p65), cRel and RelB. The proteins p50 and p52 are generated through proteolytic cleavage or arrested translation from their precursors p105 and p100, respectively[318;319]. All five members of the larger NFκB-family share an N-terminal Rel homology domain (RHD) as DNA binding domain. The members of the "Rel"-subfamily contain a further transcriptional

Introduction

activation domain at their C-terminus, whereas members of the "NFκB"-subfamily lack this activation domain, but possess instead several ankyrin domains[312]. As such, members of the "NFκB"-subfamily are generally thought not to act as activators (but can also function as repressors), except when building a heterodimer with a member of the Rel subfamily providing the activation domain. All vertebrate NFκB-family proteins have been shown to form either homo – or heterodimers *in vivo*, except RelB, which only forms heterodimers *in vitro*[312]. NFκB dimers show the common feature to bind to a highly variable 9-10 base pair long "κB" binding sequence in the human genome (5'-GGGRNWYYCC-3'; R: A/G, N: any nucleotide, W: A/T, Y: C/T). The combinatorial diversity contributes to the regulation of distinct sets of genes[312;313;317].

2.2.10.1. Regulation of NFκB

NFκB (nuclear factor kappa-light-chain-enhancer of activated B cells) is an ubiquitously expressed transcription factor and is present in most cells as a latent form in the cytoplasm, bound in a complex with IκB (inhibitor of NFκB). Thereby, binding of IκB (in the majority of cases mediated through the ankyrin domains) to NFκB covers the localization signal of the NFκB dimer and interferes with sequences involved in DNA binding, therefore retaining and inhibiting NFκB-activation[320;321]. The IκB-family consists of IκBα, IκBβ, IκBγ, IκBε,IκBζ (Bcl-3) and Cactus, each of them involved in different steps of regulation, either by phosphorylation, proteolysis, tissue specific expression or different affinities for a given NFκB dimer[322;323]. In contrast, the IκB kinase (IKK) family plays a crucial role in activating NFκB. The family consists of four members: IKKα, IKKβ, IKKε and the NFκB essential modulator (NEMO, also called IKKγ). All IKKs are serine threonine kinases that phosphorylate IκBs in a stimulus dependent manner[275]. NEMO in this context acts as a scaffold protein[324;325]. In the canonical NFκB-pathway for example, an IKK complex consisting of two NEMO molecules, IKKα and IKKβ phosphorylates IκB, leading to its degradation, followed by release and activation of NFκB (Fig. 2.5. A). On the other hand, also the inactivation of NFκB by different IKKs has been

Introduction

reported by stabilizing the IκB/NFκB interaction[326;327]. An overview of the most important molecules involved in NFκB signaling is shown in figure 2.6.

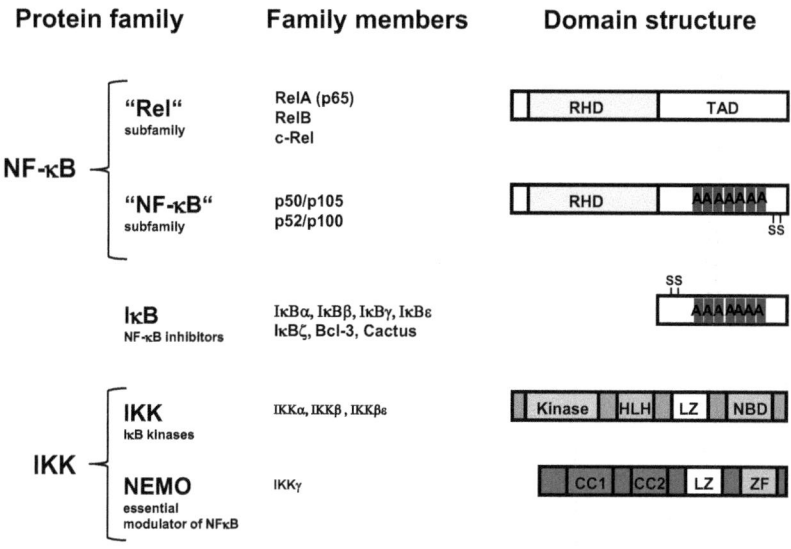

Figure 2.6. Schematic outline of the most important molecules involved in NFκB signaling (adapted from Gilmore et al.[312]) (RHD, Rel homology domain; TAD, transcriptional activation domain; A, ankyrin repeat-containing inhibitory domain; S, N-terminal serine residues; Kinase, kinase domain; HLH, helix-loop-helix; LZ, leucine zipper; NBD, NEMO binding domain; CC, coiled coil; LZ, leucine zipper; ZF, zinc finger).

2.2.10.2. NFκB-pathways

Three pathways are known for the activation of NFκB: (i) The canonical/classical pathway, (ii) the non-canonical/alternative pathway and (iii) a not yet fully understood third pathway. In the canonical NFκB-pathway (Fig. 2.7. A), NFκB dimers such as p50/RelA are retained in the cytoplasm by complex formation with an IκB molecule (in most cases IκBα). For this purpose, the binding of a ligand to its cell surface receptor (e.g. TNF to the TNF receptor or a PAMP to its Toll-like receptor) leads to the recruitment of adaptor proteins (e.g. TRAFs and RIP) to the cytoplasmatic domain of the receptor[328;329]. These

Introduction

adaptors then in turn recruit an IKK complex (consisting of IKKα and IKKβ and two molecules of the regulatory scaffold molecule NEMO) directly onto the cytoplasmic adaptors (e.g. by virtue of the K63-ubiquitin binding activity of NEMO)[324;325].

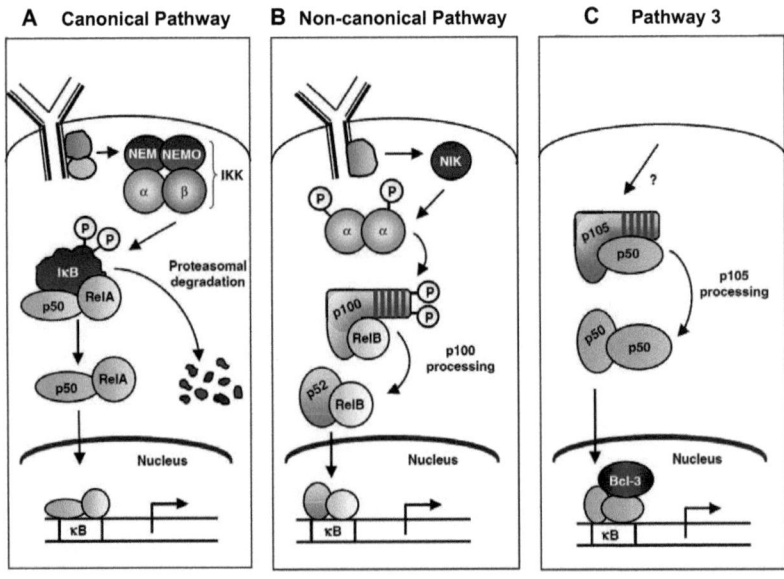

Figure 2.7. Schematic outline of NFκB-signal transduction pathways in humans (taken from Gilmore et al.[312])

This clustering of molecules at the receptor then activates the IKK complex. Afterwards, IKK phosphorylates IκB at two serine residues, leading to its ubiquitination at the position K48 and degradation by the proteasome. NFκB then translocates into the nucleus to activate different target genes. The canonical pathway is autoregulated, as NFκB activates expression of the IκBα gene that leads to resequestration of the complex in the cytoplasm by the newly synthesized IκB protein[312;330-332].

The non-canonical pathway, however (Fig. 2.7. B), mainly activates p100/RelB complexes during B and T cell organ development. Moreover, in the non canonical pathway only certain receptor signals, e.g. by Lymphotoxin B, B cell activating factor and CD40, lead to an activation of this pathway, which is processed through an IKK containing two IKKα subunits and lacking NEMO. Receptor binding results in the activation of the NFκB-inducing kinase NIK, which phosphorylates and activates an IKKα complex. This in turn phosphorylates two serine residues adjacent to the ankyrin repeat C-terminal IκB domain of p100, leading to its partial proteolysis and liberation of the p52/RelB complex[312;330-332].

In pathway 3 (Fig. 2.7. C), p50 or p52 homodimers translocate to the nucleus, where they become transcriptional activators by interaction with the IκB-like co-activator IκBζ (Bcl-3). To date, the regulation of pathway 3 is unknown[312].

2.2.10.3. The role of NFκB-transcription factors in the development of DCs

NFκB/Rel transcription factor family members are expressed at high levels in DCs[333]. The known homo- or heterodimers of the five distinct NFκB /Rel proteins (p50, p65, c-Rel, RelB and p52)[334] have all been knocked out in mice. As a result, lack of RelB or a combined deficiency of RelA and p50 or c-Rel and p50 lead to a loss of DCs[335;336]. Furthermore, RelB has been shown to play a critical role in DC maturation and immunogenicity[337-341]. To translate these results from mice into the human system, several *in vitro* studies have been performed to investigate the role of NFκB in the development and function of human moDCs. RelB for instance has been shown to regulate DC subset development[340]. Besides development, also survival of DCs has been shown to be dependent on NFκB: Inhibition of NFκB-activation by pharmacological inhibitors like caffeic acid phenethyl ester (CAPE), BAY 11-7085 (BAY) and AS602868 (AS) during and after DC differentiation resulted in induction of apoptosis[342]. Moreover, canonical NFκB activity was required for the acquisition of a DC phenotype and complete maturation[342]. Especially p65, c-Rel, and/or p50 dimers have been associated with activation-induced DC maturation and function, as the inhibition of NFκB during human DC- differentiation resulted in T cell anergy and T_{reg} activity[343-346]. Furthermore, DCs lacking proper NFκB expression were unable to exert their function as sentinels of the immune system, as TLR signaling is absolutely dependant on NFκB-activation, functionally linking antigen recognition and maturation[85;86;347]. Furthermore, cytokine[313] and maturation marker[348] expression have been shown to be controlled by NFκB signaling in human as well as in murine bone marrow derived DCs. In this context, the maturation marker CD83 has been shown to be regulated in part by NFκB: Berchtold et al. cloned and characterized the MP -261 and verified the binding of the NFκB subunits p50 and p65 to a bioinformatically predicted NFκB binding site by EMSA[167].

Taken together, the pivotal role of NFκB for DCs includes development, differentiation, maturation, functionality and cytokine as well as surface marker expression.

2.2.11. Interferon regulatory factors

Members of the IRF-family are transcription factors involved in the regulation of a variety of biological processes. Originally identified as intracellular mediators of interferon induction and activity[349], their central role in host defense to pathogens has recently been confirmed by the recognition of their involvement in the regulation of gene expression in responses to TLR triggering and other PRRs[88;89]. Playing a major role in the development as well as in the activity of hematopoietic cells puts them at the interface of innate and adaptive immune responses[350;351]. IRFs have also been reported to regulate cell growth and apoptosis in several cell types, thereby affecting susceptibility for cancer and its progression[281]. So far, nine members of the IRF-family have been identified in mice and humans based on a unique helix-turn-helix DNA binding domain at the N-terminus[281;350-352]. Two more IRF-family members, IRF-10 and IRF-11, were recently identified in birds an teleost fish, but both lack expression in mice as well as in humans and their function is yet unknown[353;354]. An overview of the human IRF-family members and their functions is given in figure 2.8.

Introduction

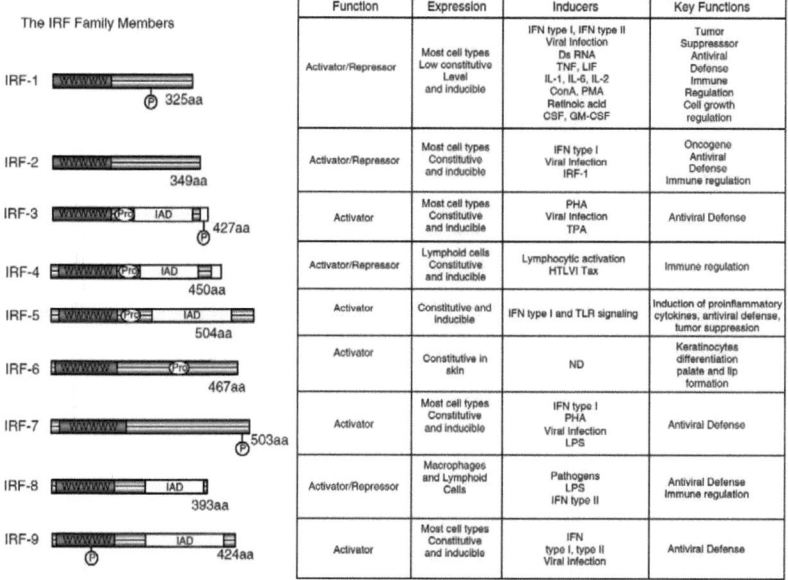

Figure 2.8. Human IRF-family members and their functions (taken from Battistini et al.[355])

2.2.11.1. Activation and regulation of IRF-pathways

Interferons were first described as virus-induced regulators of the IFN-α/β transcription through the RIG-I/MDA-5-pathway[356-361;361]. Later on their pivotal role in TLR signaling became evident, either in the MyD88-independent pathways of TLR3 and 4 through TRIF and IKKε[359-363] or the MyD88-dependent pathways of TLR7, 8 and 9 through TRAF6[363-365]. The circle was closed by the observation that IRFs are absolutely essential for the transcription of interferon stimulated genes through the JAK/STAT-pathway[366].

The activity of the interferon regulatory factors is regulated by different means: Primarily, they are regulated at the transcriptional level[367]. The expression of IRF-1 for example is regulated by several stimuli like INF-γ or TNF-α[368;369], but it is also cell cycle dependent[370;371]. Other stimuli like IL-4 reduce the INF-γ induced IRF-1 expression[372]. Furthermore, several antagonistic activation and repression domains allow a complex regulation by interaction partners like IRF-

Introduction

1 itself, IRF-2 or IRF-8, which either activate or inactivate the DNA binding and transactivation capacities of IRF-1[373;374]. Furthermore, IRF-1 autoregulates its own expression on the one hand by a negative feedback mechanism and on the other hand by the induction of e.g. IRF-2, which then in turn acts as transcriptional repressor and antagonist of IRF-1[349;375-378]. Second, the half life of the several IRF proteins differs considerably. The IRF-1 protein for example has a half life of about 30 minutes[379]. This leads to a harsh limitation of its activity, as for example negative regulating factors like IRF-2 have a longer half life of about 1-2 hours and inhibit the IRF-1 activity[379;380]. Third, phosphorylation at specific serine and threonine residues determines the DNA binding and transactivation activity of e.g. IRF-3, 4, 5 and 7. When expressed, they initially lie dormant in the cytoplasm and display an inactive DNA binding domain. Specific phosphorylation events in the cytoplasm at serine residues result in an activated phospho-IRF isoform, which translocates to the nucleus and binds then to the designated response element[329;347;381]. Especially the MyD88-independent signaling of TLR3 and 4 confers the phosphorylation of IRF-3/IRF-7 heterodimers by IKKε. Phospho-IRF-3/phospho-IRF-7 translocate subsequently to the nucleus and activate Type I IFN costimulatory molecules[88;347;350;351;381-383]. In the MyD88-dependent signaling pathway especially IRF-5 and IRF-7 are phosphorylated by TRAF6, then translocate into the nucleus and lead there to the transcription of either type I IFN (IRF-7) or inflammatory proteins (IRF-5) [88;347;350;351;381-383].

2.2.11.2. IRFs in development and function of DCs

The fundamental role of the IRF-family members for the immune system is best displayed during hematopoietic cell differentiation. Studies in both mice and humans showed that similar to NFκB, both the myeloid as well as the lymphoid cell development is dependent on several IRF family members. As displayed in figure 2.9., the antagonistic and synergistic interplay of both lineage-specific and ubiquitously expression plays a major role in lineage differentiation and maturation[384]. Especially IRF-1, IRF-2 and IRF-8 play an important role for myeloid cell development (Fig. 2.9. A)[355]. For the development, regulation, and function of lymphoid cells IRF-1, IRF-2, IRF-4, and IRF-8 are of pivotal importance (Fig. 2.9.B)[355]. However, the IRF-family plays also a crucial role especially for the development and function of DCs. Mostly studies in mice, but also in human moDCs showed the pivotal role of the IRFs for DC subset development and activation.

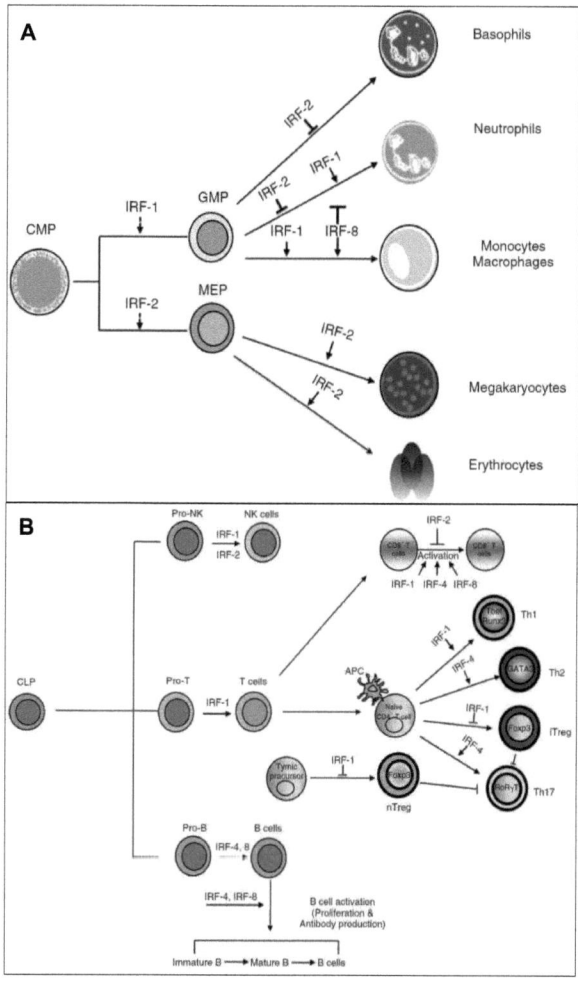

Figure 2.9. IRFs in myeloid and lymphoid cell development (taken from Battistini et al.[355])

Introduction

Some IRFs have emerged as important guideliners of DC subset development and functional diversity as they control the phenotypical development of the DC progenitor (Fig. 2.10.)[355;385]. The closely related IRF-4 and IRF-8 were first identified to have a fundamental impact on DC biology. Moreover, both TFs have been reported in mice to be expressed in a subset specific manner: IRF-8 is expressed at high levels in CD8α+ DCs and P-DCs and at low levels in CD8α− DCs[386;387]. In contrast, IRF-4 expression is high in CD4−/CD8− double negative DCs and CD4+ DCs and is marginally expressed in CD8α+ DCs and P-DCs[388]. Expression of these IRFs correlates with their requirement for DC subset development. IRF-8−/− mice have been shown to lack P-DC and CD8α+ DCs and as a consequence show impaired production of type I IFN and IL-12 p40[389]. IRF-4−/− mice, however, display a severely reduced population of CD4+ DCs[388;390]. IRF-8 is also required for full differentiation and function of epidermal Langerhans cells and dermal DCs[312]. Recent studies in IRF-1 knockout mice indicated that IRF-1 also regulates the differentiation and function of DCs thereby showing similarities to IRF-4 and IRF-8[391], as IRF-1−/− mice show reduced numbers of CD8α+ DCs. In contrast to IRF-8−/− mice, they display a predominance of P-DC differentiation. Thus, IRF-1 can also function as a negative regulator of CD4+CD8α− DC development. Interestingly, DCs from IRF-1−/− mice fail to fully mature in response to viral or bacterial stimuli and are unable to stimulate the proliferation of allogeneic T cells. Conversely, IRF-1−/− DCs represent a tolerogenic cytokine profile characterized by higher levels of TGF-β, IL-10, and of the tolerogenic enzyme IDO and induce an IL-10-mediated suppressive activity in allogeneic CD4+CD25+ regulatory T cells. Notably, IRF-1 is involved in controlling the tolerogenic features of DCs[355]. The tolerogenic characteristics of IRF-1−/− DCs also account for their constitutive suppressor mechanisms, including the maintenance of an immature or anergic DC phenotype and the induction of activated T_{reg} cells. Thus, IRF-1 is thought to be a critical element in the intimate cross talk between T_{reg} cells and tolerogenic DCs. IRF-2 also has an important role in DC development. IRF-2−/− mice exhibit a selective loss of splenic and epidermal CD4+CD8α− DCs[392;393]. Also in human DC subtypes IRF transcripts have been reported to be differentially expressed[394-396] thus affecting the development and function of DC subtypes.

Introduction

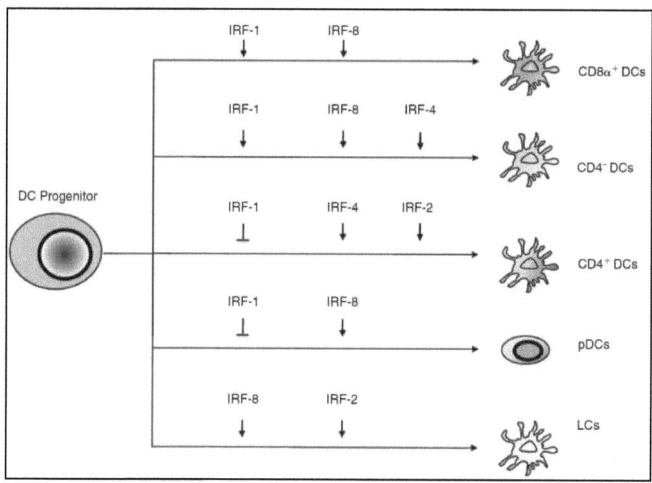

Figure 2.10. IRFs in murine DC development (taken from Battistini et al.[355])

2.2.11.3. The role of IRFs in DC activation

Toll like receptor signaling plays a pivotal role in the activation of DCs. For both mice and humans it has been shown that IRFs are activated by the MyD88-dependent and –independent pathway and different TLRs can activate distinct or shared IRFs in a cell-type- and stimulus-specific manner[88;347].

IRF-1 has also been shown to play a role in IFN induction in the MyD88-dependent signaling in myeloid DCs upon TLR9 engagement. In IFN-γ-treated cells, TLR9 triggering results in IRF-1 posttranslational modifications that allow optimal induction of a specific set of genes downstream of the TLR MyD88-dependent pathway, which are specifically upregulated by IRF-1, including IFN-β, nitric oxide synthase (iNOS), and IL-12 p35[397]. It is noteworthy that IRF-1, 3, 4, 5 and 7 can directly interact with MyD88, adding a further level to the TLR signaling complexity[397-399]. An overview of the functions of different IRFs in TLR signaling of DCs is depicted in figure 2.11.

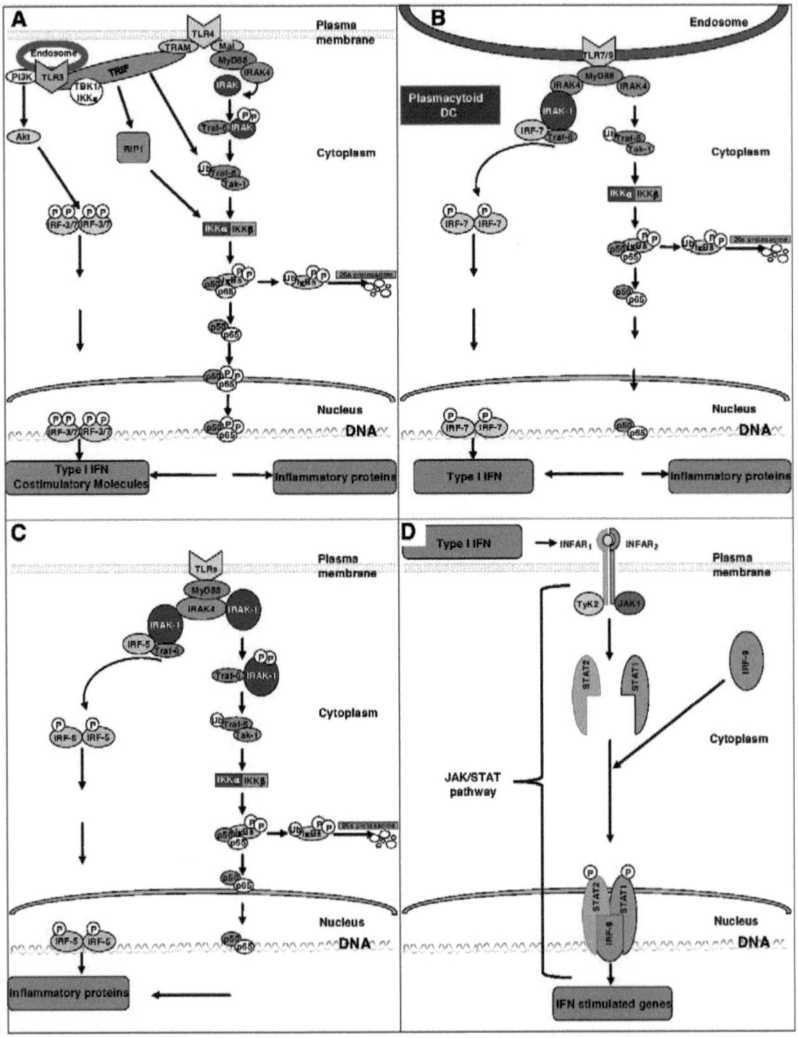

Introduction

Figure 2.11. Role of IRF family members in DC activation and TLR signaling (adapted from Moynagh[347]**).** (A) TLR3/4 signaling in plasmacytoid DCs. (B) TLR7/9 signaling in DCs. (C) General TLR signaling in DCs. (D) The JAK/STAT-pathway.

IRF-3/IRF-7 heterodimers are activated in the MyD88-independent pathway by phosphorylation upon engagement of TLR3 and TLR4 (Fig. 2.11. A)[329;400-402]. In this context, the phosphorylation of IRF-3 and -7 is mediated by the two related

Introduction

kinases, IKKε, and TBK1[403;404]. Activated IRF-3/IRF-7 heterodimers lead to the induction of type I IFN and IFN-inducible genes like the costimulatory molecules CD80, CD86 and CD40[405;406]. The induction of IFN-β by TLR4 is mainly mediated by IRF-3[364]. In the absence of IRF-3, overexpression of IRF-7 leads to the induction of IFN-β upon exposure to LPS through the MyD88-dependent pathway[402]. In this context, IRF-7 homodimers are specifically activated downstream of the murine TLR7, the human TLR8- and TLR9-pathways and lead to the induction of type I IFN (Fig. 2.11. B)[347;351;381]. IRF-7 homodimers can also be activated by the IRAK1/4-IKKα protein kinase cascade, which is known to be involved in NFκB-activation[407;408]. It has recently been reported that phosphatidylinositol-3 kinase (PI3K) is also required for optimal IRF-7 nuclear translocation and IFN production in P-DCs after TLR triggering[409]. In P-DC IRF-7 is probably the main regulator of type I IFN gene induction, as IRF-7[-/-] mice, unlike DCs from IRF-3[-/-] mice, exhibit a substantial defect in type I IFN induction following viral infection or treatment with TLR9 ligands[364]. Accordingly, these cells have a higher constitutive expression of IRF-7[395] than other cell types.

Another member of the IRF family, IRF-5, interacts with MyD88 and TRAF6 (Fig. 2.11. C) and is essential for TLR-mediated induction of proinflammatory cytokines via the MyD88-dependent pathway [382;410]. IRF-5 participates in cytokine gene expression in response to TLR7, TLR9 and TLR4 engagement, as IRF-5[-/-] mice were shown to be impaired in TLR signaling and to survive against otherwise lethal doses of CpGs or LPS[382]. IRF-5 appears to control the production of proinflammatory cytokines other than type I IFNs. Another report shows that IRF-5 regulates the induction of multiple proinflammatory cytokines in concert with NFκB by a yet-unknown MyD88-associated protein kinase[411]. Moreover, IRF-5 has been shown to be essential for MyD88-independent TLR3 signaling and is activated when TBK1 or IKKε is coexpressed[412]. Recent studies indicate a cell-type-specific role of IRF-5 in cytokine gene induction by different splice variants[397;413], but the exact mechanisms of activation and function of IRF-5 has yet to be elucidated.

As for the MyD88-dependent TLR7 signaling, IRF-4 has been shown to compete with IRF-5 for the binding to MyD88[414;398]. Indeed, IRF-4 can act as a negative regulator of TLR signaling, leading to the attenuation of IRF-5 activation and to the loss of induction of proinflammatory cytokines[347;381;398;415]. Thus, in cDCs the MyD88-dependent interplay between IRF-4 and IRF-5 may finely regulate proinflammatory cytokine levels following TLR ligation.

IRF-6, although being structurally related to IRF-5, probably does not play a role in DC function or the immune response in general, but has been reported to be the key regulator of the switch from keratinocyte proliferation to differentiation[355;381].

According to Zhao et al.[374], IRF-8 is presumably involved in TLR9 signaling. IRF-8$^{-/-}$ DCs fail to produce proinflammatory cytokines in response to CpG, but not in response to LPS[416]. In response to various PAMPs, IRF-8 is also required for the expression of IL-12 p40 and for the induction of type I IFN both in P-DCs and cDCs[417]. IFN production can be upregulated in a positive feedback loop through the enhanced expression of IRF-1,-7, and -8, which themselves are IFN-stimulated[88]. In IRF-8$^{-/-}$ DCs only the first IRF-3/7 dependent phase of type I IFN production occurs, but the second phase, which enables high IFN production, is significantly reduced. This indicates that IRF-8 may participate in the second amplifying wave of IFN production.

Finally, IRF-9 has been reported to play an important role in the IFN-induced JAK/STAT signaling (Fig. 2.11. D)[366;418;419]. It has been shown, that IRF-9 is involved in CCL19 gene expression in human moDCs[420] and the CpG-DNA-induced IFN-α production in human P-DC precursors[421].

The wide participation of IRFs in the diversity of the TLR signaling may account for the ligand- and cell-type-specific activation of target genes. Signaling from TLRs also elicits the up-regulation of co-stimulatory molecules on DCs like CD80, CD86 and CD40[405;406]. Figure 2.11. sums up the role of the most important members of the IRF-family in the MyD88-dependent and - independent pathway as well as in IFN signaling.

3. Tasks

The overall objective of this work was to fully characterize the CD83 promoter, including all regulatory elements that provide for the cell type and maturation status specificity of CD83 expression.

The first aim was to identify the genomic region containing potential regulatory elements like e.g. enhancers. This was achieved by the analyses of differential H3K9 acetylation in immature DCs (iDC), mature DCs (mDCs) and human foreskin fibroblasts (HFF) by a ChIP-chip^{TM} microarray in cooperation with Dr. I. Knippert[422]. The first 6 kb of CD83 Intron 2 were shown to be exclusively H3K9 acetylated in mDCs, hinting at a cell type and maturation status-specific hotspot of transcriptional activity in this region.

The next task was to narrow down the potential enhancer element within this region. This was achieved by mutation mutagenesis and luciferase assays. As a result, a 185 bp fragment (*185 bp enhancer*) of the H3K9 acetylated region within CD83 intron 2 has been shown to induce the CD83 minimal promoter (MP -261) in a cell type- and maturation status-specific manner in mDCs. Consequently, the next aim was to elucidate the molecular mechanism underlying the induction of the MP -261 by the *185 bp enhancer*. Therefore, a biocomputational analysis was performed in cooperation with Dr. Thomas Werner. This analysis predicted three NFκB- and five SP1-binding sites in the MP -261, two IRF- and one SP1-site in the *185 bp enhancer* and additionally two NFκB- and one IRF-site in a potential additional upstream promoter (UpP). Furthermore, the formation of a tripartite regulatory complex, consisting of UpP, MP -261 and *185 bp enhancer* was predicted. The formation of this complex was supposedly mediated by the interaction of the IRF- and NFκB-transcription factors, thereby forming three NFκB-IRF-NFκB transcriptional modules *in trans*.

To prove this model, the cooperation of all three predicted regulatory elements, namely UpP, MP -261 and *185 bp enhancer*, as well as the functionality of the predicted NFκB- and IRF-sites had to be verified. Therefore, several different experimental approaches were taken: (i) Adenoviral transduction with luciferase reporter vectors to prove the cooperation of the three regulatory elements in a chromosome-like configuration. (ii) EMSAs to prove the binding of the respective nuclear factors to the predicted binding sites. (iii) Mutation of the three IRF-sites in the luciferase reporter plasmids to verify their functionality and significance for the formation of the tripartite regulatory complex. (iv) Induction of the predicted NFκB-sites by cotransfection of plasmids coding for members of the NFκB-transcription factor family and luciferase reporter plasmids to verify their functionality. Both the transcription factor binding sites as well as the cooperation of UpP, MP -261 and the *185 bp enhancer* could be verified.

Finally, the possibility of additional epigenetic mechanisms regulating CD83 expression had to be assessed. Therefore, an analysis for cell type- and maturation status-dependent CpG methylation in the region between bp -502 to bp +212, containing both the CD83 UpP and MP -261, was performed in cooperation with *Varionostic* (Ulm). These analyses revealed that neither in iDCs, mDCs or HFF cells CpG methylation occurs at the analyzed sites and thus epigenetic mechanisms most likely are not involved in the cell type- and status-specific expression of CD83.

4. Material and Methods

4.1. Material

4.1.1. Chemicals

If not indicated differently, chemicals were purchased from *Cambrex* (Verviers, Belgium), *Invitrogen* (Karlsruhe), *Merck* (Darmstadt), *Roche* (Mannheim), *Roth* (Karlsruhe) or *Sigma-Aldrich* (Deisenhofen) in p.a. quality.

Chemicals	Chemicals
Acetic acid	Methanol (MeOH)
Acid-Citrate-Dextrose (ACD)	4-Nitrophenyl-b-D-glucuronide
29:1 40% acrylamide solution Rothiphorese	N,N,N`,N`-Tetra methyl ethylen diamine (TEMED), *Promega* (Mannheim)
Acrylamide Rothiphorese Gel 30	Nonidet P-40 (NP-40)
Agarose NEEO	Penicillin G
Ammonium persulphate (APS)	Phenol pH 4,0
Ampicillin	Phenyl methyl sulfonyl flouride (PMSF)
Bovine Serum Albumin (BSA)	Phosphoric acid (H_3PO_4)
Boric acid	Poly(dI-dC) • Poly(dI-dC) sodium salt (pIdC)
Bromophenol blue	Potassium acetate
Calcium chloride ($CaCl_2$)	Potassium chloride (KCl)
Cesium chloride (CsCl)	Potassium hydrogen carbonate (KH_2PO_4)
Chloroform	Potassium hydrogen phosphate
Chloroquine diphosphate	Propidium iodide (PI)
Deoxynucleotide mix (10mM each)	Proteinase K
D(+) Sucrose	RNAse A, *Qiagen* (Hilden)
Diethanolamine	Rotiphorese Gel 40
Dithiothreitol (DTT)	Skimmed milk powder
Diethylaminoethyl (DEAE)-Dextran	Sodium-N-lauroyl sarcosinate
Diethyl ether	Sodium acetate
Dimethyl sulphoxid (DMSO)	Sodium azide (NaN_3)
Ethanol	Sodium carbonate (Na_2CO_3)
Ethidium bromide (EtBr)	Sodium chloride (NaCl)
Ethylen diamine tetra acetate (EDTA)	Magnesium chloride ($MgCl_2$)

Materials and Methods

Formaldehyde	Magesium sulfate (MgSO$_4$ xH$_2$0)
Fetal Calf Serum (FCS)	2-Mercaptoethanol
γ-^{32}P-ATP (222TBq/mMol)	Sodium desoxycholate
Gentamycin	Sodium dodecyl sulfate (SDS)
Glycerol	Sodium fluoride (NaF)
Glycine	Sodium hydrogen carbonate (NaHCO$_3$)
Human Serum Albumin	Di-Sodium hydrogen phosphate (Na$_2$HPO$_4$)
Hydrochloric acid (HCL)	Sodium hydroxid (NaOH)
Hydrogen peroxide (H$_2$O$_2$)	Sodium pyruvate
Isopropanol	Sodium vanadate (NaVO$_3$)
Kanamycin	Sulphuric acid (H$_2$SO$_4$)
LB-Agar (Luria/ Miller)	Tetramethylbenzidine (TMB)
LB-Medium (Luria/ Miller)	Tri (hydroxymethyl) aminomethane (Tris)
L-Glutamine	Triton X-100
Lymphoprep	Trypan blue
Lithium chloride (LiCl)	Tween 20
	TEMED

Table 4.1. Chemicals

4.1.2. DNA modifying enzymes

Enzyme	Supplier
DNA restriction enzymes	New England Biolabs (NEB [Frankfurt/Main])
Alkaline Phosphatase, Calf Intestinal (CIP)	New England Biolabs (NEB [Frankfurt/Main])
T4 DNA ligase (Rapid ligation kit)	Roche (Mannheim)
Platinum Pfx polymerase	Invitrogen (Karlsruhe)
T4 Poly nucleotide kinase (T4-PNK)	New England Biolabs (NEB [Frankfurt/Main])
Klenow fragment of DNA Polymerase I	New England Biolabs (NEB [Frankfurt/Main])

Table 4.2. DNA modifying enzymes

All appropriate reaction buffers were supplied by the manufacturer.

Materials and Methods

4.1.3. Buffers and reagents

All buffers and solutions were prepared in ddH$_2$0 purified by EASYpure II ultrapure water system *(Wilhelm Werner GmbH, Leverkusen)* if not listed otherwise.

4.1.3.1. General reagents and buffers

Buffer	Ingredients
10% Sodium azide solution	10g sodium azide/100 ml
PBS (1x)	137 mM NaCl 2.7 mM KCl 4.3 mM Na$_2$HPO$_4$ x 7 H$_2$O 1.4 mM KH$_2$PO$_4$ pH 7.2
Tris-HCL pH7.5	121.14 g/l Adjust pH with HCL

Table 4.3. General reagents and buffers

4.1.3.2. Buffers and reagents for DEAE-Dextran cell transfection

Buffer	Ingredients
Chloroquine diphosphate stock solution	1 mg/ml
DEAE-Dextran stock solution	20 mg/ml DEAE-Dextran in TBS buffer
PBS/ 10% DMSO	10% DMSO in PBS
Solution A, pH 7,5	80 g/l NaCl 3.8 g/l KCl 2 g/l Na$_2$HPO$_4$ 30 g/l Tris
Solution B	15 g/l CaCl$_2$ 10 g/l MgCl$_2$
TBS buffer (1l)	100 ml Solution A 10 ml Solution B 890 ml ddH$_2$0

Table 4.4. Buffers and reagents for DEAE-Dextran cell transfection

4.1.3.3. Buffer and reagents for SDS-polyacrylamide gel electrophoresis (PAGE) and Western blot analyses

Buffer	Ingredients
Blocking buffer	5% Skimmed milk powder 0.02% Tween[20] 0.02% NaN$_3$ 10% PBS (10x)
RIPA lysis buffer	10 mM Tris-HCl pH 7.5 150 mM NaCl 1% NP-40 0.1% SDS
Running buffer (10x)	2 M Glycin 250 mM Tris-HCl 1% SDS
SDS loading dye (4x)	200 mM Tris-HCl pH 6.8 400 mM DTT 8% SDS, 12% 2-Mercaptoethanol 0.01% Bromophenol blue 40% Glycerol
Separation buffer (4x)	1.5 mM Tris-HCl pH 8.8 0.04% SDS
Stacking buffer (4x)	0.5 M Tris-HCl pH 6.8 0.4% SDS
TBST	0.02% NaN$_3$ 0.05% Tween[20] 0.5% BSA 50 mM Tris-HCl pH 7.5 150 mM NaCl
Transfer buffer (10x)	2.2 M Glycine 200 mM Tris base
Transfer buffer/ 20% MeOH (1x)	10% Transfer buffer (10x) 20% Methanol
Wash buffer	10% PBS (10x) 0.02% Tween[20]

Table 4.5. Buffer and reagents for SDS-PAGE and Western blot analyses

Materials and Methods

4.1.3.4. Reagents for chemical-competent *E.coli* and transformation by heat

Buffer/Reagent	Ingredients
0.1 M CaCl$_2$	1.11g CaCl$_2$/100 ml
0.1 M MgCl$_2$	0.952g MgCl$_2$/100 ml
LB-Agar (Luria/Miller)	40 g LB-Agar/l (supplemented with 100 mg Ampicilline/l or 50 mg Kanamycine/l when necessary)
LB-Medium (Luria/Miller)	25 g LB-Medium/l (supplemented with 100 mg Ampicilline/l or 50 mg Kanamycine/l when necessary)
Soc (super optimal broth)- medium	2% Bactotryptone pH 7 0.5% Yeast extract 10 mM NaCl 2.5 mM KCl 10 mM MgSO$_4$ 10 mM MgCl$_2$

Table 4.6. Buffers and reagents for chemical-competent *E.coli* and transformation by heat

4.1.3.5. Buffers and reagents for electrophoretic mobility shift assay (EMSA)

Buffer	Ingredients
10x TBE	890 mM Tris base 890 mM Boric acid 20 mM EDTA
10x BSB (band shift buffer)	100mM Tris-HCl (pH 7.5) 10mM DTT 50% Glycerol 10mM EDTA
1xSTE	100mM NaCl 20mM Tris-HCl (pH 7.5)
Running buffer	0.5x TBE

Table 4.7. Buffers and reagents for EMSA

Materials and Methods

4.1.3.6. Buffers and reagents for DNA electrophoresis

Buffer/Reagent	Ingredients
Agarose gel 1% (Roth, Karlsruhe)	1 g Agarose NEEO in 100 ml 1x TAE
DNA loading dye	250 ug/ml Bromophenol blue 50 mM EDTA pH 8.0 80 % (v/v) Glycerol
Ethidiumbromide solution (Roth, Karlsruhe)	10 µl 1% solution/100 ml agarose gel
Tris-Acetate-EDTA-buffer (TAE, 50x)	2 M Tris-HCl pH 8.5 5 mM Sodium acetate 62.5 mM EDTA

Table 4.7. Buffers and reagents for DNA electrophoresis

4.1.3.7. Ladders and markers

Type of ladder	Spectrum	Supplier
1 kb DNA ladder	>0.5, 1, 1.5, 2, 3, 4, 5, 6, 8, 10 kb	*NEB* (Frankfurt/ Main)
100 bp DNA ladder	> 100, 200, 300, 400, 500, 600, 700, 800, 900,1000,1200, 1517 bp	*NEB* (Frankfurt/ Main)
Full range rainbow recombinant protein standard	> 10, 15, 25, 30, 50, 75, 105, 160, 250 kDa	*Invitrogen* (Karlsruhe)
PageRuler prestained protein ladder	> 10, 15, 25, 35, 40, 55, 70, 100, 130, 170 kDa	*Fermentas* (St. Leon-Rot)
Spectra multicolor broad range protein ladder	> 10, 17, 26, 34, 42, 52, 72, 95, 135, 260 kDa	*Fermentas* (St. Leon-Rot)

Table 4.8. Ladders and markers

Materials and Methods

4.1.4. Cell culture

4.1.4.1. Cell lines and cell culture media

If not indicated otherwise, all cell culture media and reagents were purchased by Lonza (Basel, Switzerland) and PAA (Cölbe).

Cell type	Short definition	Cell culture medium
293(T)	Human embryonic kidney cell line, transformed with sheared adenovirus genome (stably transfected with large T antigen)	Dulbecco´s Modified Eagle Medium (DMEM) *(Lonza)* without L-Glutamine, 4.5 g glucose/l *(Lonza)* 10% FCS *(PAA)* 1% Penicillin/ Streptomycin/ L-Glutamine *(PAA)*
HeLa	Human cervix carcinoma cell line	Dulbecco´s Modified Eagle Medium (DMEM) *(Lonza)* without L-Glutamine, 4.5 g glucose/l *(Lonza)* 10% FCS *(PAA)* 1% Penicillin/ Streptomycin/ L-Glutamine *(PAA)*
HFF	Human foreskin fibroblast primary cells	Minimum Essential Medium (DMEM) *(Lonza)* without L-Glutamine *(Gibco)* 7.5% FCS *(PAA)* 1% L-Glutamine *(PAA)* 0.001% Gentamycin
Jurkat	T cell leukemia derived T lymphocyte cell line	RPMI1640 without L-Glutamine *(Lonza)* 10% FCS *(PAA)* 1% Penicillin/ Streptomycin/ L-Glutamine *(PAA)*
NIH3T3	BALB/c derived embryonic fibroblast cell line	Dulbecco´s Modified Eagle Medium (DMEM) *(Lonza)* without L-Glutamine, 4.5 g glucose/l *(Lonza)* 10% FCS *(PAA)* 1% Penicillin/ Streptomycin/ L-Glutamine *(PAA)*
NS47	BALB/c derived fibroblast cell line	Iscove´s Modified Dulbecco´s Medium (IMDM) *(Lonza)* without L-Glutamine *(Lonza)* 10% FCS *(PAA)* 1% Penicillin/ Streptomycin/ L-Glutamine *(PAA)* 1% Sodium pyruvate *(PAA)*

Materials and Methods

PBMC	(Human) Peripher blood mononuclear cells	RPMI1640 without L-Glutamine *(Lonza)* 1% human autologous serum 1% Penicillin/ Streptomycin/L-Glutamine *(PAA)* 1% Hepes *(Lonza)*
Raji	Burkitt's lymphoma derived lymphoblastoid B cell line	RPMI1640 without L-Glutamine *(Lonza)* 10% FCS *(PAA)* 1% Penicillin/ Streptomycin/ L-Glutamine *(PAA)* 1% Hepes *(Lonza)*
XS52	BALB/c derived DC-like cell line	Iscove´s Modified Dulbecco´s Medium (IMDM) *(Lonza)* without L-Glutamine *(Lonza)* 10% FCS *(PAA)* 1% Penicillin/ Streptomycin/ L-Glutamine *(PAA)* 1% Sodium pyruvate *(PAA)* 10% Supplement NS47 10 ng/ml murine GM-CSF
JCAM	T cell leukemia derived T lymphocyte cell line, similar to Jurkat, but lacking the LCK kinase	RPMI1640 without L-Glutamine *(Lonza)* 10% FCS *(PAA)* 1% Penicillin/ Streptomycin/ L-Glutamine *(PAA)*

Table 4.9. Cell lines and cell culture media

4.1.4.2. Additional cell culture media

Medium	Ingredients
2% FCS/PBS	2% FCS (PAA) in DPBS *(Lonza)*
Cryopreservation medium cell lines	90% FCS *(PAA)*, 10% DMSO *(Sigma-Aldrich)*
Cryopreservation medium primary cells	11% HSA *(Baxter,* Unterschleißheim*)*, 20% DMSO *(Sigma-Aldrich)* , 10% Glucose *(Roth)*
Medium	**Supplier**
Optimem I + Glutamax I	Gibco/Invitrogen (Darmstadt)

Table 4.10. Additional cell culture media

Materials and Methods

4.1.4.3. General cell culture reagents

Buffer/Reagent	Supplier
Aqua ad iniectabilia	*DeltaSelect* (München)
DPBS	*Lonza* (Köln)
Fetal calf serum	*PAA* (Pasching, Austria)
Hepes	*Lonza* (Köln)
Pen-Strep	*PAA* (Pasching, Austria)
Sodium Pyruvat	*Cambrex* (Wiesbaden)
Trypan Blue	*Cambrex* (Wiesbaden)
Trypsin EDTA	*Lonza* (Köln)

Table 4.11. General cell culture reagents

4.1.4.4. General cell culture dishes and plastic ware

All cell culture dishes and plastic ware were provided by *Greiner Bio-One* (Frickenhausen), *Nunc/Thermo Scientific* (Langenselbold) and *Millipore* (Schwalbach), if not listed otherwise.

4.1.5. Equipment and instruments

Unit of equipment	Manufacturer
Agarose gel chamber	*Peqlab* (Erlangen)
Bacteria Incubator *Incucell*	*MMM Medcenter* (Planegg / München)
Cell incubator *Hera Cell*	*Heraeus* (Hanau)
Centrifuge big *Sorvall Evolution RC*	*Thermo Scientific* (Langensebold)
Centrifuge small *Biofuge pico/fresh*	*Heraeus* (Hanau)
Electroporator *Genepulser II*	*Bio-Rad* (München)
Electroporator *Nucleofector I*	*Amaxa/Lonza* (Köln)
Fine scale *BP221S*	*Sartorius* (Göttingen)
Flowcytometer FACScan	*Becton, Dickinson and Company (BD)* (Heidelberg)
Fluorometer *Wallac Victor 2*	*Perkin-Elmer* (Rotgau)

Materials and Methods

Fume hood *MC6*	*Waldner Laboreinrichtungen* (Wangen)
Gel dryer *FSGD-4534*	*Thermo Scientific* (Langensebold)
Heatblock	*Eppendorf* (Hamburg)
Incubator shaker *Series 25*	*New Brunswick Bioscientific* (Wesseling-Berzdorf)
Laminar flow cabinet *Hera Safe*	*Heraeus* (Hanau)
Light cabinet *Alpha Imager 2200*	*Alpha Innotech Corp.* (Kasendorf)
Microwave oven *Privileg 8020E*	*Privileg* (Hamburg)
Nanodrop 2000 photometer	*Peqlab* (Erlangen)
Nitrogen freezer tank (-170°C)	*German Cryo* (Hochneukirch)
pH meter *Lab 850*	*Schott Instruments* (Mainz)
Phospho image reader *FLA-5000*	*Fuji Photo Film Co.* (Tokyo, Japan)
Power supply *Power Pac 300*	*Bio-Rad* (München)
Protein gel chamber EMSA *SE600 Ruby Complete*	*GE Healthcare* (München)
Protein gel/Western blot system *Protean II/Tetracell*	*Bio-Rad* (München)
Refrigerator (-20°C) *Liebherr Comfort*	*Liebherr* (Biberach an der Riss)
Refrigerator (-80°C) *Hera Freeze*	*Heraeus* (Hanau)
Scale *BC1500*	*Sartorius* (Göttingen)
Scintillator *TRI-Carb 2100 TR Liquid scintillation analyzer*	*Perkin-Elmer* (Rotgau)
Table-top photo processor *Curix 60*	*Agfa* (Köln)
Thermoycler *T3*	*Biometra* (Göttingen)
Ultrapure water system EASYpure II	*Wilhelm Werner GmbH* (Leverkusen)
UV table *Multilmage*	*Alpha Innotech Corp.* (Kasendorf)
Waterbath *WNB/WNE/WPE*	*Memmert* (Schwabach)

Table 4.12. Equipment and instruments

Materials and Methods

4.1.6. Software and websites

Program	Developer/Source
AIDA	Raytest
Blast/Blast2Sequence	http://blast.ncbi.nlm.nih.gov/Blast.cgi
DNA Manipulation Toolkit	http://www.vivo.colostate.edu/molkit/manip/
Excel	Microsoft
FCS Express	De Novo Software
Irfan View	Irfan Skiljan
MathInspector	Genomatix
Photoshop	Adobe
PowerPoint	Microsoft
Prism	Graphpad Software
Reference Manager	Thomson Reuters
Signal Map	Roche NimbleGen
Webcutter 2.0	http://rna.lundberg.gu.se/cutter2/
Word	Microsoft

Table 4.13. Software and websites

4.1.7. Antibodies

4.1.7.1. Antibodies used for FACS

Antibodies were purchased by BD Biosciences (Heidelberg), if not listed otherwise.

Specificity	Conjugation	Clone
Mouse anti-human CD14	PE	M5E2
Mouse anti-human CD25	PE	M-A251
Mouse anti-human CD40	PE	5C3
Mouse anti-human CD80	PE	L307.4
Mouse anti-human CD83	PE	HB15e
Mouse anti-human CD86	PE	IT2.2

Materials and Methods

Mouse anti-human HLA-A,B,C	PE	G46-2.6
Mouse anti-human HLA-DR	PE	G46.6
Mouse anti-human IgG1	PE	MOPC-21
Mouse anti-human IgG2a	PE	G155-178
Mouse anti-human IgG2b	PE	27-35

Table 4.14. Antibodies for FACS

4.1.7.2. Antibodies used for EMSA

Specificity	Clone	Company
Goat anti-human IRF-5	polyclonal	*Abcam* (Cambridge, U.K.)
Normal goat IgG	polyclonal	*Santa Cruz Biotech. Inc* (Heidelberg)
Normal rabbit IgG	polyclonal	*Santa Cruz Biotech. Inc* (Heidelberg)
Rabbit anti-human cRel	polyclonal	*Santa Cruz Biotech. Inc* (Heidelberg)
Rabbit anti-human IRF-1	polyclonal	*Santa Cruz Biotech. Inc* (Heidelberg)
Rabbit anti-human IRF-2	polyclonal	*Santa Cruz Biotech. Inc* (Heidelberg)
Rabbit anti-human IRF-3	polyclonal	*Santa Cruz Biotech. Inc* (Heidelberg)
Rabbit anti-human IRF-4	polyclonal	*Santa Cruz Biotech. Inc* (Heidelberg)
Rabbit anti-human IRF-6	polyclonal	*Santa Cruz Biotech. Inc* (Heidelberg)
Rabbit anti-human IRF-7	polyclonal	*Santa Cruz Biotech. Inc* (Heidelberg)
Rabbit anti-human IRF-8	polyclonal	*Abcam* (Cambridge, U.K.)
Rabbit anti-human p50	polyclonal	*Santa Cruz Biotech. Inc* (Heidelberg)
Rabbit anti-human p52	polyclonal	*Santa Cruz Biotech. Inc* (Heidelberg)
Rabbit anti-human p65	polyclonal	*Santa Cruz Biotech. Inc* (Heidelberg)
Rabbit anti-human RelB	polyclonal	*Santa Cruz Biotech. Inc* (Heidelberg)

Table 4.15. Antibodies for EMSA

4.1.7.3. Antibodies used for Western blot analyses

Specificity	Clone	Company
Anti-mouse IgG, HRP	polyclonal	Cell Signaling/NEB (Frankfurt a. Main)
Anti-rabbit IgG (H+L), HRP	polyclonal	Promega
Anti-rabbit IgG, HRP	polyclonal	Cell Signaling/NEB (Frankfurt a. Main)
Anti-rat IgG, HRP	polyclonal	Cell Signaling/NEB (Frankfurt a. Main)
Goat anti-human IRF-5	polyclonal	Abcam (Cambridge, U.K.)
Goat anti-human Lamin A/C	N-18	Santa Cruz Biotech. Inc (Heidelberg)
Mouse anti-human GAPDH	6C5	Millipore (Schwalbach)
Rabbit anti-goat IgG, HRP	polyclonal	DakoCytomation (Glostrup, Denmark)
Rabbit anti-human cRel	polyclonal	Santa Cruz Biotech. Inc (Heidelberg)
Rabbit anti-human IRF-1	polyclonal	Santa Cruz Biotech. Inc (Heidelberg)
Rabbit anti-human IRF-2	polyclonal	Santa Cruz Biotech. Inc (Heidelberg)
Rabbit anti-human IRF-3	polyclonal	Santa Cruz Biotech. Inc (Heidelberg)
Rabbit anti-human IRF-4	polyclonal	Santa Cruz Biotech. Inc (Heidelberg)
Rabbit anti-human IRF-6	polyclonal	Abcam (Cambridge, U.K.)
Rabbit anti-human IRF-7	polyclonal	Santa Cruz Biotech. Inc (Heidelberg)
Rabbit anti-human IRF-8	polyclonal	Abcam (Cambridge, U.K.)
Rabbit anti-human p50	polyclonal	Santa Cruz Biotech. Inc (Heidelberg)
Rabbit anti-human p52	polyclonal	Santa Cruz Biotech. Inc (Heidelberg)
Rabbit anti-human p65	polyclonal	Santa Cruz Biotech. Inc (Heidelberg)
Rabbit anti-human RelB	polyclonal	Santa Cruz Biotech. Inc (Heidelberg)
Rat anti-human CD83	polyclonal serum 1G11	GSF-Forschungszentrum, Dr. med Elisabeth Kremmer

Table 4.16. Antibodies for Western blot analyses

Materials and Methods

4.1.8. DNA Oligos

All DNA oligonucleotides and primers were purchased from MWG Eurofins Operon *(Ebersberg)*.

4.1.8.1. Primers

Notes and abbreviations:

fw, for: DNA oligonucleotides in forward orientation
rev: DNA oligonucleotides in reverse orientation
mt: mutant
wt: wild type
aa: amino acid
PCR: Polymerase chain reaction

4.1.8.1.1. Sequencing primers

Primer	Sequence	Description
5knob_for	5´-AGGCAGTTTGGCTCCAATATCTGG- 3´	Screening adenovirus genome; fw
Ad1124-1100_rev	5´-ATTTTCACTTACTGTAGACAAACAT- 3´	Screening adenovirus genome; rev
CMV fwd	5´-CGCAAATGGGCGGTAGGCGTG-3´	John Hiscott plasmid sequencing; fw
cRel 2nd run	5´-TGACATAGAAGTTCGTTTTGT-3´	cRel internal sequencing; fw
cRel 3rd run	5´-AGTGACAGCATGGGAGAGACT-3´	cRel internal sequencing; fw
E3_for	5´-CTGCTAGTTGAGCGGGACAGGGGAC-3´	Screening adenovirus genome; fw
E3_rev	5´-GGCAAGGAGGTGCTGCTGAATAAAC- 3´	Screening adenovirus genome; rev
E4_for	5´-ATTGAAGCCAATATGATAATGAGGG- 3´	Screening adenovirus genome; fw
E4_rev	5´-CACAGCGGCAGCCATAACAGTC- 3´	Screening adenovirus genome; rev
IRF-1 Test rev	5´-CATCTCCTCTTTATTAATCCAGATGAG-3´	IRF-1 internal sequencing; rev
IRF-2 1st run	5´-TAGTGTGCCCAGCGATGAAG-3´	IRF-2 internal sequencing; fw
p50 2nd run	5´-AATGGTGGAGTCTGGGAAGG-3´	p50 internal sequencing; fw
p65 2nd run	5´-GGAGTTCCATGGCTCGTCTG-3´	fw; p65 internal sequencing
p65 3rd run	5´-CTCCTGTGCGTGTCTCCA-3´	fw; p65 internal sequencing
p65 4th run	5´-TGGGTTGCACGGGATTGAAG-3´	fw; p65 internal sequencing
pcDNA3 reverse	5´-GGCAACTAGAAGGCACAGTC-3´	rev; pCDNA 3.1 sequencing
pEF fwd	5´-AGCCTCAGACAGTGGTTC-3´	fw; Kay Child plasmid sequencing

Materials and Methods

pGL reverse	5'-CTTTATGTTTTTGGCGTCTTCC-3'	rev; pGL3 sequencing
pGL3 forward	5'-CTAGCAAAATAGGCTGTCCC-3'	fw; pGL3 sequencing
SeqITR_for	5'-CGGGAAAACTGAATAAGAGGAAGTGA -3'	fw; pShuttle sequencing
Seq-Mfe-fiber_rev	5'-TGTATAAGCTATGTGGTGGTGGGG- 3'	Screening adenovirus genome
T7	5'-TAATACGACTCACTATAGGG-3'	fw; pCDNA 3.1 sequencing

Table 4.17. Sequencing primers

4.1.8.1.2. Cloning primers for transcription factors

Primer	Sequence	Description
IRF-2 fwd HindII	5'-GGCAAGCTTAACATGCCCATCACTTGG-3'	IRF-2 PCR product from Kay Childs template (aa 3, 4, 5 wrong); fw
IRF-2 rev Xbal	5'-GGCTCTAGATTAACAGCTCTTGACGCG-3'	IRF-2 PCR product from Kay Childs template (aa 3, 4, 5 wrong); rev
IRF-2mut fwd	5'-TTAAGCTTAACATGCCCGTGGAAAGGATGCGCATGCG-3'	Correction of aa 3, 4, 5; fw
IRF-2mut rev	5'-CCTCTAGATTAACAGCTCTTGACGCGGGCCTGG-3'	Correction of aa 3, 4, 5; rev
p65 HindIII fw	5'-GGCAAGCTTGCCATGGACGAACTGTTC-3'	p65 PCR product from John Hiscott template; fw
p65 Xbal rev	5'-GGCTCTAGACCCTTAGGAGCTGATCTG-3'	p65 PCR product from John Hiscott template; rev

Table 4.18. Cloning primers for transcription factors

4.1.8.1.3. Cloning primers for luciferase reporter constructs

Primer	Sequence	Description
A-1	5'-GGCGGTACCAGCTGGGGCTCTTCTCAATATTATAAAG-3'	Fragment C deletion; rev; wt
A-10	5'-GGCGGTACCAGATGATTTCCAAAGGAAGGGAC 3'	Fragment C deletion; fw; wt
A-1b	5'-GGCGGTACCCAATATTATAAAGTCTATTTATAG-3'	Fragment C deletion; 185bp enhancer; rev; wt
A-2	5'-GGCGGTACCAGGTGCCAATGGGGACAGTACG-3'	Fragment C deletion; rev; wt
A-3	5'-GGCGGTACCAGAAGGCATTGCAACTCTGG-3'	Fragment C deletion; rev; wt
A-4	5'-GGCGGTACCGATGCTTCACTCTCCTCACC-3'	Fragment C deletion; rev; wt
A-8	5'-GGCGGTACCAGTACTTTGGGCCTGGTTGATAATC-3'	Fragment C deletion; fw; wt

Materials and Methods

Primer	Sequence	Description
A-9	5´-GGCGGTACCCCTATGGGTGATGCAAAACGAAAG-3´	Fragment C deletion; 185bp enhancer; fw; wt
C-forward	5´-CCAGGGTACCGAGGAGGTATTTTGAGAAAATATG-3´	CD83 intron 2 fragment C; fw; wt
C-forward 2	5´-CCAGGGTACCACAATATCATGTCTGTGAGGAGTAAAGC-3´	Fragment C deletion; fw; wt
C-Kurz 1	5´-GGCGGTACCTATAATATTGAGAAGAGCCC-3´	Fragment C deletion; fw; wt
C-Kurz 2	5´-GGCGGTACCATTGGCACCTATAGTACTTG-3´	Fragment C deletion; fw; wt
C-Kurz 3rev	5´-GGCGGTACCCTTACGCCTGTAATCCCAGC-3´	Fragment C deletion; rev; wt
C-reverse	5´-GCAGGGTACCTTCCTCTTCTTTGTGTAGTG-3´	CD83 intron 2 fragment C; rev; wt
Intron2-A_for	5´-TTAAGGTACCGTAGGTGCTGCGATACC-3´	CD83 intron 2 fragment A; fw; wt
Intron2-A_rev	5´-CCGGGGTACCAATGAAGTAGGAATATTTAAC-3´	CD83 intron 2 fragment A; rev; wt
Intron2-B_for	5´-CATTGGTACCTTACTTACTGTGGGATCAGAG-3´	CD83 intron 2 fragment B; fw; wt
Intron2-B_rev	5´-GCAAGGTACCAAAGAACCACATCTATTACAAC-3´	CD83 intron 2 fragment B; rev; wt

Table 4.19. Cloning primers for luciferase reporter constructs

4.1.8.1.4. Cloning primers for the mutated 185 bp enhancer

Primer	Sequence	Description
185bp Enh.fw/MT	5´-GGCGGTACCCCTATGGGTGATGCAGGACTATAGAGGATATATGGTG-3´	185 bp enhancer; fw; 1st IRF-site mt
185bp Enh.fw/WT	5´-GGCGGTACCCCTATGGGTGATGCAAAACGAAAGAGGATATATGGTG-3´	185 bp enhancer; fw; wt
185bp Enh.rev/MT	5´-GGCGGTACCCAATATTATAAAGTCTATTTATAGTAGACTTTTATATGAAGTCACACTCTTATTCCTCTCCTCTCCC-3´	185 bp enhancer; rv; 2nd IRF-site mt
185bp Enh.rev/WT	5´-GGCGGTACCCAATATTATAAAGTCTATTTATAGTAGACTTTTATATGAAGTCACACTCTTACTTCCCTTTTCTCCC-3´	185 bp enhancer; rev; wt

Table 4.20. Cloning primers for the mutated 185 bp enhancer luciferase reporter constructs

Materials and Methods

4.1.8.2. Oligos for EMSA

4.1.8.2.1. Oligos NFκB-sites

Oligo	Sequence	Description
NFKB-(Pr1fw)I	5´-CGGGGACGGGGGCGCCCCGGCCTAA-3´	NFκB-site 3; enhancer; fw; wt
NFKB-(Pr1mutfw)K	5´-CGGGGACGCCGGCGACCCGGCCTAA-3´	NFκB-site 3; enhancer; fw; mt
NFKB-(Pr1mutrv)L	5´-TTAGGCCGGGTCGCCGGCGTCCCCG-3´	NFκB-site 3; enhancer; rev; mt
NFKB-(Pr1rv)J	5´-TTAGGCCGGGGCGCCCCCGTCCCCG-3´	NFκB-site 3; enhancer; rev; wt
NFKB-(Pr2fw)M	5´-GCCCGCCGGGGAATCCCCCGGGCTGG-3´	NFκB-site 4; enhancer; fw; wt
NFKB-(Pr2mutfw)O	5´-GCCCGCCGGCCAATCGGCCGGGCTGG-3´	NFκB-site 4; enhancer; fw; mt
NFKB-(Pr2mutrv)P	5´-CCAGCCCGGCCGATTGGCCGGCGGGC-3´	NFκB-site 4; enhancer; rev; mt
NFKB-(Pr2rv)N	5´-CCAGCCCGGGGGATTCCCCGGCGGGC-3´	NFκB-site 4; enhancer; rev; wt
NFKB-(Pr3fw)Q	5´-GGCGCGCAGGGAAGTTCCCGAACGC-3´	NFκB-site 5; enhancer; fw; wt
NFKB-(Pr3mutfw)S	5´-GGCGCGCACCGAAGTTGACGAACGC-3´	NFκB-site 5; enhancer; fw; mt
NFKB-(Pr3mutrv)T	5´-GCGTTCGTCAACTTCGGTGCGCGCC-3´	NFκB-site 5; enhancer; rev; mt
NFKB-(Pr3rv)R	5´-GCGTTCGGGAACTTCCCTGCGCGCC-3´	NFκB-site 5; enhancer; rev; wt
NFKB-(UP1fw)A	5´-TAAAATGGGCAGATCCCGCTCGGAT-3´	NFκB -site 1; upstream promoter; fw; wt
NFKB-(UP1mutfw)C	5´-TAAAATGCCCAGAGACCGCTCGGAT-3´	NFκB-site 1;upstream promoter; fw; mt
NFKB-(UP1mutrv)D	5´-ATCCGAGCGGTCTCTGGGCATTTTA-3´	NFκB-site 1; upstream promoter; rev; mt
NFKB-(UP1rv)B	5´-ATCCGAGCGGGATCTGCCCATTTTA-3´	NFκB-site 1;upstream promoter; rev; wt
NFKB-(UP2fw)E	5´-GCGGTCTGCAAAGCCCCCAGCGCTG-3´	NFκB-site 2; enhancer; fw; wt
NFKB-(UP2mutfw)G	5´-GCGGTCTACATAGCCACCAGCGCTG-3´	NFκB-site 2; enhancer; fw; mt
NFKB-(UP2mutrv)H	5´-CAGCGCTGGTGGCTATGTAGACCGC-3´	NFκB-site 2; enhancer; rev; mt
NFKB-(UP2rv)F	5´-CAGCGCTGGGGGCTTTGCAGACCGC-3´	NFκB-site 2; enhancer; rev; wt

Table 4.21. Oligos for EMSA (NFκB-sites)

4.1.8.2.2. Oligos IRF-sites

Oligo	Sequence	Description
A opt2	5´-ATGGGTGATGCAAAACGAAAGAGGATATATGGT-3´	IRF-site 1; enhancer; fw; wt
B opt2	5´-ACCATATATCCTCTTTCGTTTTGCATCACCCAT-3´	IRF-site 1; enhancer; rev; wt
C opt2	5´-ATGGGTGATGCAGGACTATAGAGGATATATGGT-3´	IRF-site 1; enhancer; fw; mt
D opt2	5´-ACCATATATCCTCTATAGTCCTGCATCACCCAT-3´	IRF-site 1; enhancer; rev; mt

Materials and Methods

E opt	5´-AAGGAAGGGAGAAAAGGGAAGTAAGAGTGTGAC-3´	IRF-site 2; enhancer; fw; wt
F opt	5´-GTCACACTCTTACTTCCCTTTTCTCCCTTCCTT-3´	IRF-site 2; enhancer; rev; wt
G opt	5´-AAGGAAGGGAGAGGAGAGGAATAAGAGTGTGAC-3´	IRF-site 2; enhancer; fw; mt
H opt	5´-GTCACACTCTTATTCCTCTCCTCTCCCTTCCTT-3´	IRF-site 2; enhancer; rev; mt
I opt	5´-GCCCGGTTCCCGGCTTCCTTTTGCGGGTCAACG-3´	IRF-site 3; upstream promoter; fw; wt
J opt	5´-CGTTGACCCGCAAAAGGAAGCCGGGAACCGGGC-3´	IRF-site 3;upstream promoter; rev; wt
K opt	5´-GCCCGGGTCCCGGCTACCTGCTGCGGGTCAACG-3´	IRF-site 3;upstream promoter; fw; mt
L opt	5´-CGTTGACCCGCAGCAGGTAGCCGGGACCCGGGC-3´	IRF-site 3; upstream promoter; rev; mt

Table 4.22. Oligos for EMSA (IRF-sites)

4.1.9. Plasmid vectors

4.1.9.1. Donated or purchased vector

Plasmid list
CMVßL/cRel (AG Stürzl and John Hiscott; Department of Molecular and Experimental Surgery; University Hospital Erlangen, Erlangen)
CMVßL/p50 (AG Stürzl and John Hiscott; Department of Molecular and Experimental Surgery; University Hospital Erlangen, Erlangen)
CMVßL/p65 (AG Stürzl and John Hiscott; Department of Molecular and Experimental Surgery; University Hospital Erlangen, Erlangen)
EF1α/IRF-2/ (aa 3, 4, 5 defective) (Kay Childs, Division of Basic Medical Sciences St. George's, University of London, U.K.)
GENEART standard vector/CD83_-510 (GENEART, Regensburg)
GENEART standard vector/CD83_-510mut (GENEART, Regensburg)
pAdEasy™ XL Adenoviral Vector System (Stratagene/Agilent, Waldbronn)
pCDNA 3.1 (-)/IRF-1 (Martin Schmidt, Department of Virology; University Hospital Erlangen)
pCDNA 3.1/IRF-5 (Martin Schmidt, Department of Virology; University Hospital Erlangen)
pEGFP-N1 (Clontech, Heidelberg)
pGL3/Basic (Promega, Mannheim)
pGL3/CMV_luc (D.M. Nettelbeck, DKFZ & Department of Dermatology; University Hospital Heidelberg)
pGL3/MP-261 (Susanne Berchtold, Department of Dermatology; University Hospital Erlangen)
pGL3/SV40 (Promega, Mannheim)

Materials and Methods

pUC 19 (*Invitrogen*, Karlsruhe)
RP1-258E1 (Human Genome Mapping Project (HGMP) Resource Centre, Hinxton, Cambridge, U.K.)
RP3-380E11 (Human Genome Mapping Project (HGMP) Resource Centre, Hinxton, Cambridge,U.K.)

Table 4.23. Donated or purchased vectors

4.1.9.2. Cloned vectors

4.1.9.2.1. Luciferase reporter constructs

Notes and abbreviations:

s: sense
as: antisense
CDS: coding sequence
enh.: enhancer
UPP: upstream promoter
S1: spacer sequence 1

Donor/template	Excision method	Insert	Recipient	Result
pEGFP-N1	Restriction digest; HindIII/XbaI	GFP CDS	pGL3/CMV_luc; HindIII/XbaI (luc CDS replaced)	pGL3/CMV_GFP
RP1-258E1	PCR; KpnI	CD83 intron 2 fragment A	pGL3/MP-261; KpnI	pGL3/fragment A s_as/MP-261
RP1-258E1	PCR; KpnI	CD83 intron 2 fragment B	pGL3/MP-261; KpnI	pGL3/fragment B s_as/MP-261
CMVßL/cRel	Restriction digest; Hind III/XbaI	cRel CDS	pCDNA 3.1; Hind III/XbaI	pCDNA 3.1/cRel
CMVßL/p50	Restriction digest; Hind III/XbaI	p50 CDS	pCDNA 3.1; Hind III/XbaI	pCDNA 3.1/p50
CMVßL/p65	PCR; Hind III/XbaI	P65 CDS	pCDNA 3.1; HindIII/XbaI	pCDNA 3.1/p65
EF1α/IRF-2/	PCR; HindIII/XbaI	IRF-2 CDS; aa 3, 4, 5 from IRF-1!	pCDNA 3.1; HindIII/XbaI	pCDNA 3.1/IRF-2;

Materials and Methods

pCDNA 3.1/IRF-2	PCR; HindIII/XbaI; aa 3, 4, 5 corrected	IRF-2 CDS corrected	pCDNA 3.1; HindIII/XbaI	pCDNA 3.1/IRF-2corrected
pGL3/185bp enh. s/MP-261	Restriction digest; KpnI	185 bp enhancer	pGL3/Basic; KpnI	pGL3/185bp enh. s_as
pGL3/fragment C/MP-261	PCR; KpnI	185 bp enhancer	pGL3/MP-261; KpnI	pGL3/185bp enh s_as /MP-261
GENEART standard vector/CD83_-510	Restriction digest; NheI/SmaI (blunt)	UPP	pGL3/Basic; NheI/XhoI (blunt)	pGL3/UPP
GENEART standard vector/CD83_-510	Restriction digest; NheI/SmaI (blunt)	UPP	pGL3/185bp enh s_as /MP-261; NheI/XhoI (blunt)	pGL3/185bp enh. s_as/UPP
RP3-380E11	PCR; SacI/MluI	S1	pGL3/UPP SacI/MluI	pGL3/UPP+S1
RP3-380E11	PCR; SacI/MluI	S1	pGL3/185bp enh. s_as/UPP SacI/MluI	pGL3/185bp enh. s_as/UPP+S1
GENEART standard vector/CD83_-510	Restriction digest; NheI/XhoI	P -510	pGL3/MP-261 NheI/XhoI	pGL3/P -510
GENEART standard vector/CD83_-510	Restriction digest; NheI/XhoI	P -510	pGL3/185bp enh s_as /MP-261 NheI/XhoI	pGL3/185bp enh. s_as/P -510
RP3-380E11	PCR; SacI/MluI	S1	pGL3/P -510; SacI/MluI	pGL3/P -510+S1
RP3-380E11	PCR; SacI/MluI	S1	pGL3/185bp enh. s_as/P -510; SacI/MluI	pGL3/185bp enh. s_as/P -510+S1
RP3-380E11	PCR; SacI/MluI	S1	pGL3/MP-261; SacI/MluI	pGL3/MP-261+S1
RP3-380E11	PCR; SacI/MluI	S1	pGL3/185bp enh s_as /MP-261; SacI/MluI	pGL3/185bp enh s_as /MP-261+S1
RP3-380E11	PCR; SacI/MluI	S1	pGL3/185bp enh s_as; SacI/MluI	pGL3/185bp enh s_as+S1
RP3-380E11	PCR; KpnI	CD83 intron 2 fragment C	pGL3/MP-261; KpnI	pGL3/fragment C s_as/MP-261
RP3-380E11	PCR; KpnI	CD83 intron 2 fragment A	pGL3/MP-261; KpnI	pGL3/fragment A s_as/MP-261
RP3-380E11	PCR; KpnI	CD83 intron 2 fragment B	pGL3/MP-261; KpnI	pGL3/fragment B s_as/MP-261
pGL3/fragment C/MP-261	PCR; KpnI	CD83 intron 2 fragments C1-C14 (s. table 4.26.)	pGL3/MP-261; KpnI	pGL3/fragment C 1-C14 s_as/MP-261

Table 4.24. Cloned luciferase reporter constructs

Materials and Methods

4.1.9.2.2. Luciferase reporter constructs with mutated IRF-sites

Notes and abbreviations:

IRF-Sites 1 and 2 in the *185 bp enhancer* were mutated by PCR mutagenesis. IRF-site 3 in the P -510 was mutated by GENEART and subcloned in the "GENEART standard vector "CD83_-510mut 3rd IRF-Site". Combinations of the 3 IRF-site mutations resulted in constructs, bearing 1, 2 or 3 mutated IRF-sites:

mut: mutated
pGL3/ UpP 3.IRFmut
pGL3/ P-510 3.IRFmut
pGL3/E.s_as/P-510 1.IRFmut
pGL3/E.s_as/P-510 2.IRFmut
pGL3/E.s_as/P-510 3.IRFmut   These constructs were subsequently extended with the spacer sequence "S1"
pGL3/E.s_as/P-510 1.+2.IRFmut
pGL3/E.s_as/P-510 1.+2.+3.IRFmut

Donor/template	Excision method	Insert	Recipient	Result
GENEART standard vector/CD83_-510mut 3rd IRF-Site	Restriction digest; NheI/XhoI	P -510mut 3rd IRF-Site mut	pGL3/MP-261 NheI/XhoI	pGL3/ P-510 3.IRFmut
GENEART standard vector/CD83_-510mut 3rd IRF-Site	Restriction digest; NheI/SmaI (blunt)	UPPmut 3rd IRF-Site mut	pGL3/Basic; NheI/XhoI (blunt)	pGL3/ UpP 3.IRFmut
pGL3/185bp enh. s/MP-261	PCR; KpnI	185 bp enhancer 1st IRF-site mut	pGL3/P -510; KpnI	pGL3/E.s_as/P-510 1.IRFmut
pGL3/185bp enh. s/MP-261	PCR; KpnI	185 bp enhancer 2nd IRF-site mut	pGL3/P -510; KpnI	pGL3/E.s_as/P-510 2.IRFmut
pGL3/185bp enh. s/MP-261	PCR; KpnI	185 bp enhancer $1^{st}/2^{nd}$IRF-site mut	pGL3/P -510; KpnI	pGL3/E.s_as/P-510 1.+2.IRFmut
pGL3/185bp enh s/MP-261	PCR; KpnI	185 bp enhancer $1^{st}/2^{nd}$IRF-site mut	pGL3/P -510mut 3rd IRF-Site; KpnI	pGL3/E.s_as/P-510 1.+2.+3.IRFmut

Materials and Methods

RP3-380E11	PCR; SacI/MluI	S1	pGL3/ P-510 3.IRFmut; SacI/MluI	pGL3/S1/P-510 3.IRFmut
RP3-380E11	PCR; SacI/MluI	S1	pGL3/E.s_as/P-510 1.IRFmut; SacI/MluI	pGL3/E.s_as/S1/P-510 1.IRFmut
RP3-380E11	PCR; SacI/MluI	S1	pGL3/E.s_as/P-510 1.+2.IRFmut; SacI/MluI	pGL3/E.s_as/S1/P-510 1.+2.IRFmut
RP3-380E11	PCR; SacI/MluI	S1	pGL3/E.s_as/P-510 1.+2.+3.IRFmut; SacI/MluI	pGL3/E.s_as/S1/P-510 1.+2.+3.IRFmut

Table 4.25. Cloned luciferase reporter constructs with mutated IRF-sites

Materials and Methods

4.1.9.2.3. Schematic depiction of CD83 Intron 2 fragments for luciferase assay

Intron 2 was divided into 3 fragments (A, B, C) and subcloned into the pGL3/MP-261 reporter construct. Subsequently fragment C was narrowed down via deletion mutagenesis in the fragments C1-C14.

Fragment of CD83 intron 2	Forward primer	Reverse primer
Fragment A 1239 bp	Intron2-A_for	Intron2-A_rev
Fragment B 2359 bp	Intron2-B_for	Intron2-B_rev
Fragment C 1-2220 bp	C-forward	C-reverse
Fragment C1 bp 1-1720	C-forward	C-Kurz 3rev
Fragment C2 bp 1-1010	C-forward	A-2
Fragment C3 bp 1-525	C-forward	A-1
Fragment C4 bp 500-2220	C-Kurz 1	C-reverse
Fragment C5 bp 500-1720	C-Kurz 1	C-Kurz 3rev
Fragment C6 bp 1000-1720	C-Kurz 2	C-Kurz 3rev
Fragment C7 bp 100-510	C-forward 2	A-1b
Fragment C8 bp 1-405	C-forward	A-4
Fragment C9 bp 100-405	C-forward 2	A-4
Fragment C10 bp 1-300	C-forward	A-3
Fragment C11 bp 100-300	C-forward 2	A-3
Fragment C12 bp 225-510	A-8	A-1b
Fragment C13 bp 325-510	A-9	A-1b
Fragment C14 bp 425-510	A-10	A-1b

Table 4.26. Overview of luciferase reporter constructs and PCR primers used for cloning fragments of CD 83 intron 2

Materials and Methods

4.1.10. Adenoviruses

Adenovirus	Specification
Ad5TL	➢ Replication deficient adenovirus serotype 5; E1 region replaced by a CMV-GFP cassette ➢ Kindly provided by D.T. Curiel, Birmingham, AL, USA
Ad5Luc1	➢ Replication deficient adenovirus serotype 5; E1 region replaced by a CMV-luciferase cassette ➢ Kindly provided by D.T. Curiel, Birmingham, AL, USA
Ad261/S1	➢ Replication deficient adenovirus serotype 5; E1 region replaced by a MP -261/S1-luciferase cassette
Ad510/S1	➢ Replication deficient adenovirus serotype 5; E1 region replaced by a P -510/S1-luciferase cassette
Ad261/S1+Es	➢ Replication deficient adenovirus serotype 5; E1 region replaced by a MP -261/S1/185 bp enhancer sense-luciferase cassette
Ad261/S1+Eas	➢ Replication deficient adenovirus serotype 5; E1 region replaced by a MP -261/S1/185 bp enhancer antisense-luciferase cassette
Ad510/S1+Es	➢ Replication deficient adenovirus serotype 5; E1 region replaced by a P -510/S1/185 bp enhancer sense-luciferase cassette
Ad510/S1+Eas	➢ Replication deficient adenovirus serotype 5; E1 region replaced by a P -510/S1/185 bp enhancer antisense-luciferase cassette
AdBasic/S1+Es	➢ Replication deficient adenovirus serotype 5; E1 region replaced by a 185 bp enhancer sense-luciferase cassette
AdBasic/S1+Eas	➢ Replication deficient adenovirus serotype 5; E1 region replaced by a 185 bp enhancer sense-luciferase cassette

Table 4.27. Adenoviruses

4.1.11. Human cytokines and maturation agents

Cytokine/ Agent	Stock concentration	Purity	Company
Lipopolysaccharide (LPS)	1 mg/ml		Sigma-Aldrich, München
Recombinant GM-CSF	4×10^4 U/ml	›95%	Cell Genix, Freiburg
Recombinant IL-4	2×10^5 U ml	›95%	Cell Genix, Freiburg
Recombinant IL-6	1×10^6 U/ml	›95%	Cell Genix, Freiburg

Table 4.28. Human cytokines and maturation agents

4.1.12. Bacteria

DH5α "Maximum Efficiency", *Invitrogen* (Darmstadt)
> F-ɸ80 Δ lacZ Δ M15, recA1, endA1, hsdR17, PhoA, supE44, gyrA96, relA1

4.1.13. Purchased kits

Kit	Tool	Company
EndoFree® Plasmid Maxi Kit	DNA purification from bacteria culture	Qiagen, Hilden
Lipofectamine/PLUS/LTX	Transfection of adherent cells	*Invitrogen, Darmstadt*
Luciferase Assay System Freezer Pack	Luciferase reporter assay	*Promega, Mannheim*
Nuclear/Cytosol Fractionation Kit	Nuclear/Cytosol fractionation	*BioCat, Heidelberg*
Nucleo Spin Tissue	DNA purification from tissue	*Macherey-Nagel, Düren*
NucleoSpin Plasmid	DNA purification from bacteria culture	*Macherey-Nagel, Düren*
Pierce BCA Protein Assay Kit	Protein concentration determination	*Thermo Scientific, Schwerte*
QIAamp DNA Mini kit	DNA purification from cells	*Qiagen, Hilden*
QIAquick® Gel Extraction Kit	Agarose gel extraction	*Qiagen, Hilden*
QIAquick® PCR Purification Kit	PCR purification	*Qiagen, Hilden*
QIAshredder	RNA purification from cells	*Qiagen, Hilden*
Rapid DNA Ligation Kit	DNA ligation	*Roche, Mannheim*
RNeasy® Mini Kit	RNA purification from cells	*Qiagen, Hilden*

Table 4.29. Purchased kits

4.2. Methods

4.2.1. General molecular biology methods

4.2.1.1. Generation of chemical-competent *E.coli* and transformation by heat

A single E. coli colony of DH5α "Maximum Efficiency" (*Invitrogen*) was inoculated in 5 ml LB medium and shaken overnight at 37°C. The next day, this pre-culture was transferred into 300 ml LB medium in a 2 l-Erlenmeyer flask and further cultivated at 37°C to an OD_{600} of 0.8. Bacteria were then transferred to 50 ml PP-tubes (*Greiner*) and put on ice for 15 min. After centrifugation at 3000 rpm for 5 min at 4°C and removal of supernatant, each bacterial pellet was resuspended in 25 ml of 0.1M $MgCl_2$ and incubated on ice for 30 min. Bacteria were again harvested at 3000 rpm, 5 min at 4°C and supernatant was removed. Each pellet was then resolved in 2 ml of 0.1 M $CaCl_2$ and incubated on ice for an additional hour. Afterwards, glycerol (final concentration of 20%) was added to the bacterial suspension and aliquots of 200 µl were frozen at -80°C.

For transformation, competent bacteria were thawed on ice and 100 µl of bacteria were mixed with either 1 µg of plasmid DNA or with 20 µl of DNA from a ligation reaction and then incubated on ice for 20 min. Afterwards, the bacteria-DNA-suspension was heated at 42°C for 2 min and then immediately put on ice for 5 min. Next, 1 ml of pre-warmed Soc-medium was added to the transformed bacteria. Bacteria were shaken for 60 min at 37°C, scratched out on LB-agar plates containing the required antibiotic and incubated at 37°C overnight.

Materials and Methods

4.2.2. DNA based molecular biology methods

4.2.2.1. Isolation of DNA

4.2.2.1.1. Isolation of small plasmids (< 12 kb)

Depending on the amount of bacterial lysate, plasmids up to a size of 12 kb were isolated either with the Plasmid Mini- or Maxi Preparation Kit from *Machery-Nagel* or *Quiagen*. Therefore a single colony of transformed bacteria was inoculated either in 4 ml or 150 ml of LB medium containing the required antibiotic and DNA was prepared according to the manufacturer's instructions. In brief, these protocols are based on a modified alkaline lysis procedure, followed by binding of plasmid DNA to anion-exchange resin under appropriate low-salt conditions. Plasmid DNA is eluted in a high-salt buffer and then desalted by isopropanol precipitation. To avoid LPS-contamination during transfection of cells, only plasmid DNA was used for this purpose which was prepared with the EndoFree Maxi Kit from *Qiagen*.

4.2.2.1.2. Isolation of large plasmids (> 12 kb)

For the isolation of small amounts of large plasmids (" Mini Preparation"), like AdEasy vectors, buffers from the purchased kits from *Qiagen* were used; however, DNA preparation was modified as follows: A colony of transformed bacteria was grown at 30°C for approx. 20 h in 5 ml LB medium containing the required antibiotic. Afterwards bacteria were harvested (600 rpm, 1min) in a 2 ml reaction-tube, resuspended in 200 µl buffer P1 (*Qiagen*) and lysed by careful mixing with 400 µl buffer P2 (*Qiagen*). The reaction was stopped after an incubation period of 5 min at RT by gently mixing with 300 µl buffer P3 (*Qiagen*). The lysate was centrifuged at 13000 rpm for 10 min in a table-top centrifuge and supernatant was then transferred to a fresh 1.5 ml reaction tube. This centrifugation step was repeated and the resulting supernatant was mixed with 750 µl of isopropanol to precipitate DNA. DNA was pelleted by centrifugation at 13000 rpm for 10 min followed by a washing step with 1 ml of

Materials and Methods

70% EtOH. DNA was air-dried for some minutes and finally resolved in 30 µl buffer TE, pH8 (*Qiagen*).

For isolation of large plasmids with the Midi Preparation Kit from *Qiagen* (Hilden) a single colony of transformed bacteria was inoculated in 100 ml of LB medium containing the required antibiotic and grown overnight at 30°C. DNA was prepared according to the manufacturer's instructions; however, the indicated buffer-volumes were increased: 10 ml buffer P1, 20 ml buffer P2, and 15 ml buffer P3. Finally, plasmid DNA was eluted in a high-salt buffer and then desalted by isopropanol precipitation.

4.2.2.1.3. Isolation of human genomic DNA from cells by QIAamp DNA Mini Kit

Chromosomal DNA was used as a template for bisulfate/pyrosequencing based DNA methylation analysis. DNA was prepared using the QIAamp Mini Kit (*Quiagen*) according to the manufacturer's instructions. Purity and concentration of the genomic DNA was assessed by optical density (see 4.2.2.2.).

4.2.2.1.4. Preparation of DNA fragments from PCR reactions and enzymatic digestions

For the purification of DNA fragments derived from enzymatic reactions or PCR-products from primers, nucleotides, polymerases and salts, the QIAquick PCR Purifikation Kit (*Qiagen*) was used according to the manufacturer's instructions.

4.2.2.1.5. Preparation of DNA fragments from gel electrophoresis

DNA fragments derived from enzymatic reactions, which were afterwards used for ligation reactions, were extracted and purified from standard agarose gels with the QIAquick Gel Extraction Kit from *Qiagen* according to the manufacturer's instructions.

4.2.2.2. Nucleic acid quantification

The optical density (OD) of either DNA or RNA was determined with a spectrophotometer (*NanoDrop 2000c, Peqlab*) by the measurement of the extinction at λ= 260 nm. To calculate DNA or RNA concentration, OD absorbance values ($A°_{260}$) were multiplied by 50 or 40, respectively, assuming $1.0A° = 50$ ng/µl dsDNA and $1.0A° = 40$ ng/µl RNA.

4.2.2.3. Separation of DNA or RNA using agarose gel electrophoresis

Agarose gel electrophoresis was used to separate nucleic acids by size in a constant isoelectric field. After separation in a gel containing 1-3% agarose in TAE buffer, DNA or RNA was visualized under UV light by the fluorescent dye ethidium bromide. The size of DNA fragments was determined by comparison with a standard DNA-ladder (*NEB*) of 100 bp or 1 kb.

4.2.2.4. Polymerase chain reaction (PCR) based methods

4.2.2.4.1. Single-step PCRs

PCR was used to (i) amplify DNA fragments for cloning from either chromosomal- or plasmid DNA and (ii) to screen adenoviral DNA for insertions.

(i) DNA fragments amplified for cloning were generated in a final reaction volume of 50 µl using a DNA polymerase containing proof-reading activity (Platinum Pfx DNA Polymerase, *Invitrogen*) according to the manufacturer's instructions.

(ii) PCR reactions for screening of adenoviral DNA were performed in a final reaction volume of 25 µl with a recombinant Taq DNA Polymerase from *Invitrogen* according to the manufacturer's instructions.

Materials and Methods

Step	Temperature	Time	Cycles
Initial denaturation	95°C	5 min	1x
Denaturation	95°C	1 min	25x
Annealing	56°C - 62°C	1 min	to
Extension	68°C	0.5 min - 2 min	30x
Final extension	68°C	10 min	1x
Pause	4°C	∞	

Table 4.30. PCR program for amplification of DNA fragments

Step	Temperature	Time	Cycles
Denaturation	95°C	5 min	1x
Denaturation	95°C	1 min	25x
Annealing	56°C - 62°C	1 min	to
Extension	72°C	1min - 2 min	40x
Final extension	72°C	10 min	1x
Pause	4°C	∞	

Table 4.31. PCR program for screening genomic DNA

Finally, 10 µl (i) to 16 µl (ii) of each reaction were analyzed by agarose gel electrophoresis and the rest of (i) was used for DNA purification (see 4.2.2.1.5.).

4.2.2.5. Hybridization of DNA oligos for electrophoretic mobility shift assays (EMSA)

For binding of transcription factors and radioactive labeling for EMSA double stranded oligos coding for the putative transcription factor binding site were needed. Therefore the single stranded oligos in sense and antisense orientation were mixed in equal molar amounts and hybridized in a thermo cycler reaction.

Reaction mix
6,4 µl H_2O
6 µl oligo „sense"
6 µl oligo "antisense"
1.6 µl 5 M NaCl
Final volume 20 µl

Table 4.32. Reaction mix for hybridization of DNA oligos for EMSA

Materials and Methods

The reaction mix was incubated in a thermo cycler *(Biometra)* using the following program:

Hybridization reaction
Lid heater 95° C
95°C 10min
Decrease temperature in 5° C Steps for 15 min. each down to 40°C
Pause 37°C forever

Table 4.33. Program for hybridization of DNA oligos for EMSA

The next day 20 µl of the reaction mix containing the hybridized oligos were diluted with 180 µl of H_2O to a final concentration of 3 pMol and then immediately frozen to -20°C until further usage. Hybridization was assessed by agarose gel electrophoresis in a 3% agarose gel.

4.2.2.6. Radioactive labeling of DNA oligos for EMSA

In order to detect the DNA oligos in EMSA, they were labeled radioactively with γ-^{32}P-ATP (222Bq/mMol) (FP301, *Hartmann Analytic*) using the T4 polynucleotide kinase *(NEB)*. The reaction was incubated at 37°C for 60 min. in a heatblock *(Eppendorf)*. The labeling reaction mix was prepared as follows:

Reaction mix
19 µl H_2O
2 µl hybridized oligos (3 pMol/µl)
3 µl 10x PNK buffer (NEB)
2 µl T4-PNK (NEB)
4 µl γ-^{32}P-ATP (222/Bq/mMol) (FP301, Hartmann Analytic)
Final volume 30 µl

Table 4.34. Reaction mix for radioactive labeling of DNA oligos for EMSA

Afterwards 30 µl of the reaction mix were diluted with 40 µl of STE reaching a final volume of 70 µl. Subsequently, residual unbound γ-^{32}P-ATP was removed

Materials and Methods

from the reaction mix by purification with ProbeQuant G50 Micro Columns (*Ge Healthcare*) according to manufacturer's instructions. In brief: Columns were vortexed to homogenize the gel matrix. The tip was broken off and the columns were centrifuged at 700 g for 2 min in a microcentrifuge (*Eppendorf*) at RT. The radioactively labeled oligos were applied onto the columns and again centrifuged at 700 g for 2 min. Thereby, the flow through was collected in 1.5 ml reaction tubes. Subsequently, labeling and purification of the oligos was assessed by measuring radioactive counts per minute (cpm) of the oligo. Therefore, 2 µl of the radioactive oligo were diluted in 5 ml of Ultima Gold Cocktail (*Perkin-Elmer*) and measured in a liquid scintillator (*Perkin-Elmer*). The oligos were subsequently diluted to a final concentration of 20000 cpm/µl in 1x STE. Diluted oligos were then immediately frozen at -20°C until further usage in the EMSA reaction.

4.2.2.7. Cloning of DNA fragments or PCR products

For the cloning of different plasmids either fragments of DNA derived from other plasmids by enzymatic digestion, or PCR products were used.

4.2.2.7.1. Dephosphorylation of cleaved DNA

To avoid religation of cleaved DNA fragments, dephosphorylation was performed after enzymatic digestion. Therefore 2 µl of Alkaline Phosphatase (CIP, *NEB*) was added to the reaction and was incubated at 37°C for 1 h.

4.2.2.7.2. Conversion of DNA overhangs

To generate blunt ends on DNA-restriction-fragments with sticky ends, Klenow fragment of DNA Polymerase I, which contains a 3´→5´ exonuclease activity, was used to fill in recessed 3´ ends of DNA fragments with 5´ overhang and to digest protruding 3´ overhangs. Cleaved DNA fragments were incubated with 2.5 µl dNTP mix (4 mM each; *Invitrogen*), 1 µl buffer 2 (*NEB*), 5 µl distilled water and 1.5 µl Klenow (*NEB*) and incubated at 25°C in a thermomixer (*Eppendorf*) for 1 h.

4.2.2.7.3. Ligation

Ligation of insert-DNA into vector-DNA was performed at a volume ratio of 9:1 in a final volume of 21 µl. Ligation reaction contained additionally 10 µl 2x ligation buffer and 1 µl T4-DNA-Ligase (both Rapid DNA Ligation Kit, *Roche*) and was incubated 5 min at RT. Subsequently the whole reaction mix was transformed into competent bacteria (see 4.2.1.1.).

4.2.2.7.4. Sequencing of DNA

To exclude mutations in cloned PCR-derived inserts or wrong insertions, all cloned vector-plasmids were sequenced by *Eurofins MWG Operon* (Ebersberg).

4.2.2.7.5. Synthesis of DNA sequences

DNA sequences that were not amplified by PCR or excised from already existing templates were synthesized by *GENEART* (Regensburg) and provided in the GENEART standard vector for further cloning.

4.2.3. Protein biochemistry methods

4.2.3.1. Generation of cell lysates

4.2.3.1.1. Preparation of whole cell extracts for SDS-polyacrylamide gel electrophoresis (PAGE)

Whole cell extracts for SDS-PAGE were generated as follows: Cells were washed 1x with ice-cold DPBS (Dulbecco´s PBS, *Lonza*) and then resuspended in an appropriate amount of RIPA lysis buffer supplemented with 0.1 M $NaVO_3$, 1 M NaF and 0.1 M PMSF (in MeOH) (20 µl/ml each). Lysates were incubated for at least 1 h at 4°C before centrifugation at 13000 rpm, 4°C for 30 min. Afterwards supernatants were transferred to a fresh reaction tube and immediately stored at -20°C.

Materials and Methods

4.2.3.1.2. Preparation of nuclear extracts for EMSA and SDS-PAGE

In order to minimize unspecific binding of cytosolic proteins to the oligos in the EMSA, the nuclei of immature DCs (iDC), mature DCs (mDC) and human foreskin fibroblasts (HFF) were isolated and lysed. Therefore, a nuclear/cytosolic fractionation was performed with the "Nuclear/Cytosol Fractionation Kit" *(BioCat)*. In brief: According to the manufacturer's instructions, 10×10^6 cells were harvested and washed 1x with DPBS. Cells were then collected by centrifugation (500 g, 5 min, and 4°C) and cells were resuspendet by vortexing in 1 ml of CEB-A (cellular extraction buffer A) supplemented with 1 µl DTT and 2 µl proteinase inhibitor mix (both supplied with the Nuclear/Cytosol Fractionation Kit). Cell suspension was then incubated for 10 min on ice. Next, 55 µl of CEB-B (cellular extraction buffer B) was added, the reaction mix vortexed and incubated for 1 min on ice. Subsequently the suspension was vortexed again and centrifuged at 16000 g for 5 min and 4°C to pellet the nuclei. The supernatant (cytosolic fraction) was removed and immediately stored at -80 °C. The resulting pellet was resuspended in 100 µl of NEB (nuclear extraction buffer), supplemented with 0.1 µl DTT and 0.2 µl proteinase inhibitor mix and incubated on ice for 40 min. During incubation the nuclei suspension was vortexed vigorously every 10 min to aid nuclear membrane lysis. Afterwards the lysate was centrifuged for 10 min at 16000 g and 4°C. The supernatant (nuclear fraction) was immediately portioned into 20 µl aliquots and stored at -80°C until further usage. CEB-A, CEB-B and NEB reagents are supplied with the Nuclear/Cytosol Fractionation Kit. Protein concentrations of cytosolic and nuclear fractions were determined by BCA Protein Assay reagent (see 4.2.3.1.3.). The purity of the fractions was assessed by Western blot analyses (see 4.2.3.3.) using control antibodies reacting to Lamin A/C and GAPDH.

4.2.3.1.3. BCA Protein Assay Reagent (bicinchoninic acid)

To determine protein concentrations of cell extracts for SDS-PAGE, Western blot analyses and EMSA, aliquots of albumin standard (*Thermo Scientific*) in different concentrations (2.0 µg/µl, 1.0 µg/µl, 0.5 µg/µl, 0.25 µg/µl, 0.125 µg/µl, 0.0625 µg/µl, 0.03125 µg/µl, 0.0 µg/µl) as well as 25 µl of cell lysate or 5 µl cell

Materials and Methods

lysate diluted in 25 µl distilled water were mixed with 200 µl working solution (*Thermo Scientific*, Pierce BCA Reagent A and B 50:1 ratio) according to the manufacturer's instructions. Finally absorption was measured with a "Wallac Victor^{2}" reader (*Perkin-Elmer*) at λ= 570 nm and protein concentrations were determined by the standard curve in a self-written Excel-program.

4.2.3.2. EMSA

In order to determine the binding of transcription factors to a radioactively labeled DNA oligo, an EMSA was performed. Thereby the binding could be shown as a bandshift in a polyacrylamide gel (see 4.2.3.2.2.). Further, the addition of a specific antibody (see table 4.15.) to the binding reaction led to a supershift of the corresponding binding protein, if it is recognized by the antibody. The addition of 200x molar excess of not labeled probe led to a "*cold*" competition, thus the loss of the bandshift signal, as a specificity control.

4.2.3.2.1. EMSA bandshift and supershift reaction

The bandshift/supershift reaction mix was prepared as follows:

Reaction mix
3 µl pldC (1 µg/µl)
1.5µl 10xBSB
2.86 µl nuclear extract (10 µg lysate diluted in 0.42 M NaCl)
add 13 µl H$_2$O
2 µl labeled oligo (20000 cpm/µl)
Final volume 15 µl

Table 4.35. Reaction mix for EMSA bandshift and supershift reaction

To reduce the background signal, 1 µg/µl of Poly(dI-dC) • Poly(dI-dC) sodium salt solution (*Sigma-Aldrich*) was used as a blocking agent. In case of a super shift or a *cold* competition, the binding reaction mix was set up without the labeled oligo, but including 5 µg of the antibody or 200x molar excess of the *cold* competition oligo. Samples were then incubated on ice for 45 min to allow

Materials and Methods

antibody or competition binding, before 2 µl of the labeled probe were added. In either case the reaction mix containing the labeled oligo was again incubated on ice for 45 min to facilitate protein-DNA binding.

4.2.3.2.2. EMSA non denaturizing polyacrylamide gel run and signal detection

To visualize the binding of a protein to a given DNA oligonucleotide, radioactively labeled DNA oligos were incubated with nuclear cell extracts and separated by a 6.66% polyacrylamide gel under non denaturing conditions.

The 6.66% polyacrylamide gel was prepared as follows:

Polyacrylamide gel mix
10 ml 29:1 40% acrylamide solution (Rotiphorese, Roth)
3 ml 10xTBE
add 47 ml H_2O
500 µl 10% APS
50 µl TEMED
Final volume 60.55 ml

Table 4.36. Polyacrylamide gel mix for EMSA

After the assembly of the gel chamber and preparation of the 6.66% polyacrylamide gel mixture, the polymerized gel was pre-run for 1 h in 0.5x TBE buffer at 160 V before loading of the binding reaction mix. One lane was loaded with a 6x protein loading dye (*Fermentas*) to assess the dye front. After the gel run (4/5 of gel range for loading dye) the gel was transferred onto a 3 mm Whatman paper (*Whatman*, Dassel) and sealed with household foil. The sealed gel was dried for at least 1.5 h on a gel dryer (*Perkins-Elmer*) by a vacuum air pump at 85°C. Afterwards the sealed gel was transferred into an imaging cassette and a BAS-MS2040 imaging plate (*Fuji Photo Film Co.*, Düsseldorf) was applied. Gel and imaging plate were incubated 1-5 days. After incubation the imaging plate was read out using a FLA-5000 phospho imaging scanner (*Fuji Photo Film Co.*) at 635 nm and a resolution of 50 µm. Data were subsequently visualized and analyzed by the AIDA software (*Raytest*, Straubenhardt).

4.2.3.3. Discontinuous SDS-PAGE: Laemmli method

Under denaturing conditions this one-dimensional gel electrophoresis separates proteins based on molecular size as they move through a polyacrylamide gel matrix towards the anode. The gel was cast as a 10% (2.25 ml separating buffer, 3.75 ml water, 3 ml acrylamide, 60 µl 10% APS, and 12.5 µl TEMED) or 12.5% (2.25 ml separation buffer, 3 ml water, 3.75 ml acrylamide, 60 µl 10% APS, and 12.5 µl TEMED) separating gel, topped by a stacking gel (2.1 ml stacking buffer, 5.15 ml water, 1.1 ml acrylamide, 60 µl 10% APS, and 17.5 µl TEMED) and was secured in an electrophoresis apparatus. Sample proteins were prepared by boiling in the presence of a loading dye mixed with SDS and 2-mercaptoethanol for 10 min at 95°C. Afterwards 20 µg of total protein (for each sample) was loaded onto the gel, which was initially run at 80 V until samples entered the separating gel and then at 120 V until the bromophenol blue-band of the loading dye reached the end of the separating gel.

4.2.3.4. Western blot analyses

Proteins separated by SDS-PAGE were transferred onto nitrocellulose filters (*Schleicher & Schuell*) with a pore size of 0.2 µm with the wet blotting device "Mini-Protean II Cell and System" (*BioRad*) at 400 mA for 45 min. Subsequently, the filters were blocked with blocking buffer for 1 h at RT to prevent unspecific binding of the antibody. After three successive washes with distilled water, primary antibody diluted 1:1000-1:100 in TBST was added and incubated for 1 h (RT) or overnight (4°C). Thereafter, filters were washed three times for 5 min with PBS/0.02% Tween[20] before they were incubated for 1 h at 4°C with the corresponding secondary horseradish peroxidase (HRP)-conjugated anti-mouse, anti-rabbit, or anti-rat IgG antibody diluted 1:10000-1:2000 in PBS/0.02% Tween[20]/5% skimmed milk powder. Finally, the filters were washed three times for 20 min with PBS/0.02% Tween[20] before the antibody/enzyme-complexes were visualized by enhanced chemiluminescence (ECL, *PIERCE*) according to the manufacturer's instructions.

4.2.4. Cell culture

4.2.4.1. Generation of dendritic cells (DC)

Human monocyte-derived DCs (moDC) were generated as described by *Berger et al.* (2002). In short, PBMC were prepared from leukapheresis products of healthy donors (obtained following informed consent and approved by the institutional review board) by density centrifugation using Lymphoprep (*Axis-Shield PoC AS*, Oslo, Norway). PBMC were resuspended in autologous cell culture medium and cells were seeded in T175 tissue culture flasks (*EasyFlasks^{TM} Nunc*, Wiesbaden) at a density of 3.5 x 10^8 cells/ flask and incubated for 1 h at 37°C in a humidified atmosphere of 5% CO_2 to allow for adherence. Subsequently, the non adherent cell fraction (NAF) was removed and cryopreserved (see 4.2.4.2.), while 35 ml of RPMI1640 cell culture medium (*Lonza*) supplemented with 250 U/ml recombinant human IL-4, 800 U/ml recombinant human GM-CSF (both *CellGenix*, Freiburg) and 1% autologous serum was added to the adherent cells (day 1). After 72h (day 4) 5 ml of fresh cell culture medium containing recombinant human GM-CSF and recombinant human IL-4 was added to the cells to the final cytokine concentrations of 250 U/ml for IL-4 and 400 U/ml for GM-CSF. The next day (day 5), cells (as immature moDCs) were used for electroporation (see 4.2.6.2.3.), adenoviral transduction (see 4.2.7.5.), lysed for whole cell lysates (see 4.2.3.1.1.) or used for preparation of nuclear extracts (see 4.2.3.1.2.). If matured cells were required, LPS was added on day 5 directly to the cell culture medium to a final concentration of 0.1 ng/ml and cultured for 24 h at 37°C in a humidified atmosphere of 5% CO_2. For maturation after infection or electroporation, LPS was added 4 h after the experimental treatment. On day 6 cells were used as mature moDCs for the subsequent experiments. The generation and maturation status of DCs were assessed by antibody staining and subsequent FACS analysis (see 4.2.5.).

Materials and Methods

4.2.4.2. Cryopreservation of primary cells

Cells of the non adherent fraction (NAF), PBMCs as well as iDCs were resuspended each in 2 ml of 20% HSA (*Baxter*, Unterschleißheim) at a concentration of 3×10^8 /ml and stored for 10 min on ice. An equal volume of cryopreservation medium consisting of 11% HSA, 20% DMSO (*Sigma-Aldrich*, Deisenhofen) and 10% Glucose (*Merck*, Darmstadt) was added to the cell suspension, mixed carefully and then frozen at -80°C.

4.2.4.3. Cryopreservation of cell lines

Cells derived from cell lines were grown in a T75 tissue culture flask (*EasyFlasks^{TM} Nunc*, Wiesbaden), harvested, washed once with ice-cold DPBS, resuspended in 6 ml of cryopreservation medium consisting of 90% FCS (*PAA*) and 10% DMSO (*Sigma-Aldrich*, Deisenhofen), portioned to 6x 1.5 ml-cryopreservation tubes and then immediately frozen at -80°C before transferred to liquid nitrogen.

4.2.4.4. Heat inactivation of fetal calf serum (FCS) and human autologous serum

In order to inactivate complement components, FCS was incubated for 30 min at 56°C in a waterbath (*Memmert*) before being added to the cell culture medium. Human autologous serum was centrifuged at 3000 rpm for 10 min in a *Big Evolution* centrifuge (*Sorvall*) and the supernatant filtered through a 0.22 µm sterile filter (Millipore) before being added to the cell culture medium.

4.2.5. Flow cytometric analysis (FACS)

For detection of cell surface molecules or GFP expression, cells were washed and resuspended at 1×10^5 cells in 100 µl cold FACS solution (DPBS containing 2% FCS) and incubated with labeled mAB, the appropriate isotype controls or left untreated for 30 min at 4°C in the dark. Afterward, cells were washed twice and resuspended in 100 µl cold FACS solution with or without 1 µg propidium iodide (*Roth*) and then analyzed with a FACScan cell analyzer

(*BD Biosciences*). Cell debris was eliminated from the analysis using a gate on forward and side light scatter. An additional gate was set on the living cells (i.e. negative for propidium iodide). A minimum of 10^4 living cells was analyzed for each sample and results were analyzed using FCS Express software (*De Novo Software*).

4.2.6. Transient transfection methods and luciferase reporter assay

4.2.6.1. DNA-transfection using the DEAE-Dextran method

For promoter analyses, adherent XS52, NIH3T3 and HeLa cells were transfected with reporter plasmid DNA containing the promoter constructs and a gene for the firefly luciferase by the DEAE-Dextran method in triplicates. Hence, 2 x 10^5 cells per well were seeded in 12-well tissue culture plate (*Falcon*) and grown overnight at 37°C, 5% CO_2. The next day, 2.5 µg of endotoxin-free plasmid DNA were diluted in 150 µl of TBS buffer and mixed with 50 µl 5 mg/ml DEAE-Dextran solution reaching a final volume of 200 µl per transfection. Subsequently 0.2 µl chloroquine solution (1 µg/µl, *Roth*) were added to the reaction mix. Cells were washed 1x with 500 µl/well TBS buffer, the DNA-DEAE-Dextran-solution was added and plates were incubated on a rocker for 30 min to 1 h at RT. Afterwards, the DNA-DEAE-Dextran-solution was replaced by 500 µl of 10% DMSO for 2 min, cells were washed with 2 x 2 ml warm DPBS and finally 2 ml per well of warm cell culture medium was added. Transfected cells were cultured for 2 days at 37°C in a humidified atmosphere of 5% CO_2 before luciferase reporter assays (see 4.2.6.5.) were performed. Transfection efficiency was assessed by transfection of a GFP coding control plasmid and subsequent FACS analyses (see 4.2.5.).

4.2.6.2. Electroporation of Raji and Jurkat cell lines

For promoter analyses suspensions of Raji and Jurkat cells were transfected by electroporation with reporter plasmid DNA containing the promoter construct and a gene for the firefly luciferase. Therefore, 10 x 10^6 cells were harvested, washed 1x in DPBS and resuspended in 250 µl fetal calf serum (PAA) containing 20 µg of DNA. Cell suspensions were incubated 10 min at RT. Subsequently 250 µl of RPMI1640 without additives were added and cells were transferred to a 4 mm electroporation cuvette (Peqlab). Electroporation was performed with a Genepulser II (Bio-Rad) at 975 µF and 260 Volt. Cells were incubated for 3 min at RT and then transferred into 10 ml prewarmed cell culture medium and cultivated at 37°C, 5% CO_2. After 24 h 10 ml of cell culture medium were added. Cells were then analyzed using a luciferase assay (see 4.2.6.5.) 48 h after electroporation. The electroporation efficiency was assessed by the electroporation of a GFP coding control plasmid and subsequent FACS analysis (see 4.2.5.).

4.2.6.3. Lipofection of DNA with Lipofectamine^{TM} LTX and PLUS^{TM} (Invitrogen) reagent

For the generation and assembly of adenoviruses, 293 cells were transfected with Lipofectamine^{TM} LTX without PLUS^{TM} reagent in a T25 tissue culture flask (Nunc) as follows: For optimal transfection efficiency, cells were transfected at 40-60% confluency. On the day of transfection, 4 µg of PacI-digested pAd-plasmid in 20 µl distilled water was mixed with 13 µl Lipofectamine^{TM} reagent in 500 µl OptiMEM (Gibco) and incubated for 30 min at RT. Cells were washed 1x with DPBS and covered with 2.5 ml OptiMEM. After the incubation the transfection mix was added directly to the cells and incubated for 6 h at 37°C in a humidified atmosphere of 5% CO_2. Afterwards the mixture was replaced by 7 ml of warm cell culture medium and cells were grown for 10-12 days until viral plaques could be detected. Viruses were then further amplified as described in 4.2.7.

For cotransfection of luciferase reporter constructs and pCDNA3.1 vectors coding for transcription factors, the Lipofectamine^{TM} LTX with PLUS^{TM} reagent was used. Therefore, 6 x 10^4 293T cells per well were seeded in a 24-well tissue culture plate (Falcon) one day prior to transfection in 500 µl antibiotic free cell culture medium. On the day of transfection 0.5 µg of total DNA, consisting of 0.05 µg reporter construct, 0.15 µg for each transcription factor construct and pCDNA 3.1 vector backbone to fill up to 0.5 µg, were diluted in 100 µl OptiMEM

Materials and Methods

(*Gibco*). Then 0.5 µl per transfection mix PLUS reagent were added to the transfection mix, mixed vigorously and incubated for 5 min at RT. Subsequently, 1.25 µl Lipofektamin^{TM} LTX was added to the transfection mix, vortexed and incubated 30 min at RT. After incubation the whole reaction mix was added drop wise directly to the cells. Transfected cells were cultured for 2 days at 37°C in a humidified atmosphere of 5% CO_2 before luciferase reporter assays (see 4.2.6.5.) were performed. Cell culture medium was changed after 24 h. The transfection efficiency was assessed by the lipofection of a GFP coding control plasmid and subsequent FACS analyses (see 4.2.5.).

4.2.6.4. Electroporation of DCs with DNA using AMAXA technology

For further promoter analyses iDC were electroporated using the AMAXA Human Dendritic Cell Nucleofector Kit (*Lonza*) and the Nucleofector I electroporation device (*Lonza*), according to the manufacturer's instructions. In brief: After harvesting and washing with warm DPBS, 2 x 10^6 iDCs were resuspended in 100 µl freshly prepared electroporation solution (provided with the Nucleofector Kit) containing 4 µg of plasmid DNA. Cells were transferred to the electroporation cuvette and electroporated with program "*U-2*". Immediately after electroporation 500 µl of RPMI1640 without additives were added to the cuvette and the whole cell suspension was transferred in 12-well tissue culture plate (*Falcon*) containing 600 µl of prewarmed RPMI1640 supplemented with 2% autologous serum, 500 U/ml human recombinant IL-4 and 800 U/ml human recombinant GM-CSF (both *CellGenix*). Directly after the transfer, cell suspension was divided into 2 wells by pipetting 600 µl in an empty well. After 4 h 1.4 ml RPMI1640 supplemented with 1% autologous serum, 250 U/ml IL-4, 400 U/ml GM-CSF and, for cells to be matured, with LPS to a final concentration of 0.1 ng/ml, was added to the cells, which were cultured 24 h at 37°C in a humidified atmosphere of 5% CO_2 before luciferase reporter assays (see 4.2.6.5.) were performed. The electroporation efficiency, generation and maturation status of the DCs were assessed by the electroporation of a GFP coding control plasmid and antibody staining, respectively and subsequent FACS analyses (see 4.2.5.).

Materials and Methods

4.2.6.5. Luciferase reporter assay

Determination of luciferase activity of DNA-transfected or adenovirally transduced cell lines and moDCs were performed in triplicates with the Luciferase Assay System (*Promega*) 1-2 days after transfection. Adherent 293T, HeLa, XS52 and NIH3T3 cells were washed with warm DPBS and then 200 µl/well 1x Luciferase Cell Culture Lysis Reagent (*Promega*, diluted 1:5 with distilled water) were added before plates were frozen for at least 2 h to -80°C. Raji, Jurkat, JCAM and DCs were first transferred to a reaction vial and then washed in 1 ml warm DPBS. Afterwards cells were collected by centrifugation (500g, 5 min, RT) and lysed with 200 µl/vial 1x Luciferase Cell Culture Lysis Reagent and frozen at -80°C for at least 2 h. Subsequently, plates or vials were thawed at RT and 10 µl (293T cells) or 20 µl (DC, Raji and Jurkat, JCAM) of cell lysate were mixed with 50 µl of Luciferase Assay Substrate (*Promega*) in a 96-well LumiNunc plate (*Nunc*). Determination of RLUs (relative luminescence units) was performed in a Wallac fluorometer (*Perkin-Elmer*, Rotgau) and normalized to the protein concentration of the lysate.

4.2.7. Recombinant adenoviruses

4.2.7.1. Cloning of plasmids containing the recombinant adenoviral genome

For the generation of plasmids containing the recombinant adenoviral genome the pAdEasy1-system (*He et al., 1998*) was used. Resulting plasmids were transfected into 293 cells for virus assembly and amplification.

4.2.7.2. Preparation of recombinant adenoviruses

All viruses were amplified in 293 cells and purified by two rounds of CsCl equilibrium density gradient ultracentrifugation. Therefore, transfected 293 cells which show viral plaques (see 4.2.8.2.) were harvested, centrifuged at 1100 rpm for 5 min at 4°C and then resuspended in 5 ml RPMI1640/2% FCS. Virus was released from cells by 3 freeze-thaw cycles and cell debris was removed by centrifugation of the lysate at 4000 rpm for 15 min at 4°C. Supernatant (2 ml) containing the virus was used to infect new 293 cells in increasing numbers (up to 15x T175 flasks) to amplify the virus; the remaining supernatant was stored at -80°C for further rounds of virus production. Viruses from 15x T175 flasks were prepared by 3 freeze-thaw cycles as described above and then loaded onto a CsCl gradient. For the CsCl equilibrium density gradient, 3 ml of CsCl at a density of 1.41 g/ml was overlaid with 5 ml of CsCl at a density of 1.27 g/ml and then the supernatant containing the enriched virus was filled up to 7 ml with DPBS and loaded onto the gradient. After ultracentrifugation at 32000 rpm for 2 h at 4°C, the virus band was harvested, diluted in HEPES buffer (8 ml) and loaded onto a second CsCl equilibrium density gradient as described above. Viruses were spun in an ultracentrifuge at 32000 rpm for 24 h at 4°C, before harvesting and purification with a PD-10 tip (*GE Healthcare*) according to the manufacturer's instructions. Finally, eluted virus (in DPBS) was mixed with 10% glycerin and aliquots of 25 µl were frozen at -80°C. Verification of viral genomes and exclusion of wild-type contamination was performed by PCR (see 4.2.2.5.1.). Physical particle concentration [viral particles (vp)/ml] was determined by OD_{260} reading and infectious particle concentration was determined by $TCID_{50}$ assay using 293 cells.

4.2.7.3. Determination of the physical particle concentration

The physical particle concentration is defined by the amount of virus particles per ml [vp/ml]. To quantify the viral DNA, the virus preparation was diluted at different ratios (1:3, 1:5, 1:10, 1:50, 1:100) with viral lysis buffer (VLB [10 mM TE, 0.5% SDS]) and incubated at 56°C in a thermomixer (*Eppendorf*) for 10 min. Cooled samples were measured with a spectrophotometer (*Eppendorf*) at OD_{260} and the mean value was calculated considering the dilution factors. Physical particle concentration was determined using the following formula:

physical particle concentration = mean value x 1.1×10^{11} vp/ml

Materials and Methods

4.2.7.4. Determination of the infectious particle concentration

The concentration of infectious virus particles has been determined using the "tissue culture infectious dose 50" ($TCID_{50}$)- method. Therefore, 293 cells were seeded in a 96-well tissue culture plate (*Falcon*) at a density of 10^5/well in RPMI1640/2% FCS. Viruses were diluted from 10^{-1} to 10^{-12} in RPMI1640/2% FCS and cells were infected with two dilution series each. Cells were incubated at 37°C, 5% CO_2 and 10 as well as 12 days after infection cell lysis was assessed microscopically. Wells showing at least one plaque were considered "positive". Infectious particle concentration was calculated using the following formula:

$$\text{Infectious particle concentration} = 10^{1+d(s-0.5)} \, TCID_{50} \, /ml$$

d = Log of sum positive wells per dilution
s = number of positive wells

4.2.7.5. Adenoviral transduction of cells

Immature day 5 DCs were seeded in 12-well tissue culture plates (*Falcon*) at a concentration of 1 x 10^6 cells/well in 250 µl cell culture medium supplemented with 800 U/ml GM-CSF and 500 U/ml IL-4. Then, adenovirus at a $TCID_{50}$ of 500/cell in a final volume of 250 µl cell culture medium without cytokines was added to the cells. After 1.5 h of incubation at RT, 2.5 ml of cell culture medium replenished with cytokines as described before was added per well. If mDCs were used, LPS was added 4 h after transduction to a final concentration of 0.1ng/ml.

Raji, Jurkat and JCAM cells were seeded in 12-well tissue culture plates (*Falcon*) at a concentration of 1 x 10^6 cells/well in 250 µl RPMI1640 cell culture medium supplemented with 2% FCS. Then, adenovirus at a $TCID_{50}$ of 500/cell (Jurkat and JCAM cells) or 50 (Raji) in a final volume of 250 µl RPMI1640 cell culture medium supplemented with 2% FCS was added to the cells. After 1.5 h of incubation at RT on the rocker, 2.5 ml of cell culture medium replenished with 10% FCS were added per well. To determine transduction efficacy, cells were transduced with Ad5TL and the percentage of living green fluorescent cells was assessed by flow cytometric analysis with a "FACScan" flowcytometer (*BD Biosciences*, Heidelberg). Only experiments that yielded transduction efficiencies of more than 70% were evaluated and are shown.

4.3. Statistical analysis

Statistical analyses were performed using the Prism software *(Graphpad)*. P-values for multiple comparisons were determined via one way ANOVA and Bonferroni's Multiple Comparison *post hoc* test. P-values for the comparison of two data sets were determined via Student's T-test and the Bonferroni-Holm *post hoc* test.

5. Results

The aim of this work was to fully characterize the human CD83 promoter, including also regulatory elements that contribute to the cell type and maturation status specificity in human dendritic cells (DC). Immature DCs (iDCs) strongly upregulate CD83 on their surface during maturation. Human CD83 is a 45kDa glycoprotein expressed predominantly on mature DCs (mDCs) and is to date one of the best known surface markers for mDCs. In 2002 Berchtold et al. identified the 261 bp CD83 minimal promoter (MP -261)[167]. They reported the isolation of a 3037 bp fragment upstream of the CD83 translation initiation codon. Subsequently, using deletion mutagenesis, a 261 bp fragment that displayed the highest promoter activity has been identified. The MP -261 was shown to be inducible by TNF-α and contained four SP1 transcription factor binding sites and one NF-κB element that were mandatory to exert the highest promoter activity[167]. However, despite being active in luciferase reporter assays, the MP -261 showed neither a cell type- nor a maturation status-specific activity. Thus, additional regulatory elements were proposed, which could be characterized in the present thesis work.

5.1. Previous approaches to identify the human CD83 promoter

5.1.1. CD83 is not expressed in HFF cells, weakly in iDCs and strongly in mDCs.

In order to establish an appropriate cell system for the monitoring of cell type- and maturation status-specific CD83 expression, human foreskin fibroblasts (HFF), monocyte derived human iDCs and monocyte derived human mDCs were analyzed for CD83 expression.

To follow CD83 expression in these cells, surface expression of the human CD83 protein was assessed by flow cytometric analyses (FACS). Therefore, iDCs were either matured for 20 hours (h) with LPS or left untreated and then harvested at the same time as the HFF cells. Next, cells were stained with a PE-labeled antibody against human CD83 as well as with the corresponding

Results

isotype control and subsequently analyzed by FACS. As displayed in figure 5.1. (A), flow cytometric analyses revealed that HFF cells did not express CD83 on their surface, whereas iDCs expressed very low and mDCs very high amounts of CD83 in comparison to the isotype control.

Next, CD83 protein expression was assessed by Western blot analyses using whole cell lysates generated from iDCs, maturing DCs and HFF cells (Fig. 5.1. B). Therefore, iDCs were either left untreated (referred to as "0h" time point) or matured with LPS for 3 h, 6 h, 10 h and 24 h. The untreated iDCs as well as HFF cells served as a negative control. Whole cell lysates from HFF displayed no signal for CD83 expression. Immature DCs displayed a weak CD83 signal at the expected molecular weight of approximately 45 kDa that gained in intensity over the maturation time of 3-, 6-, and 10 h and reached its peak at 24 h. The GAPDH loading control (~36 kDa molecular weight) showed the loading of equal amounts of protein per lane.

Third, CD83 mRNA expression in HFF cells, iDCs and 20 h LPS-matured DCs was detected by real time PCR performed by Ilka Knippertz[422] (Fig. 5.1. C). Therefore, total RNA was isolated from iDCs, mDCs and HFF cells and reversely transcribed using an Oligo (dT)$_{12\text{-}18}$ primer. Afterwards, PCR was performed with an intron spanning primer pair for CD83 and a control primer pair for GAPDH. As negative control a PCR reaction was performed where the DNA template was replaced by water, referred to as "Water" control in figure 5.1. (C). HFF cells showed no CD83 mRNA expression, whereas iDCs showed weak and mDCs a strong CD83 mRNA expression pattern.

Taken together, these data confirm that HFF cell do not express CD83 on mRNA or protein level, whereas iDCs show a weak and mature DCs a strong CD83 expression on both mRNA and protein level. Hence, these three cell types provide an optimal basis for the analyses of differential CD83 expression patterns.

Results

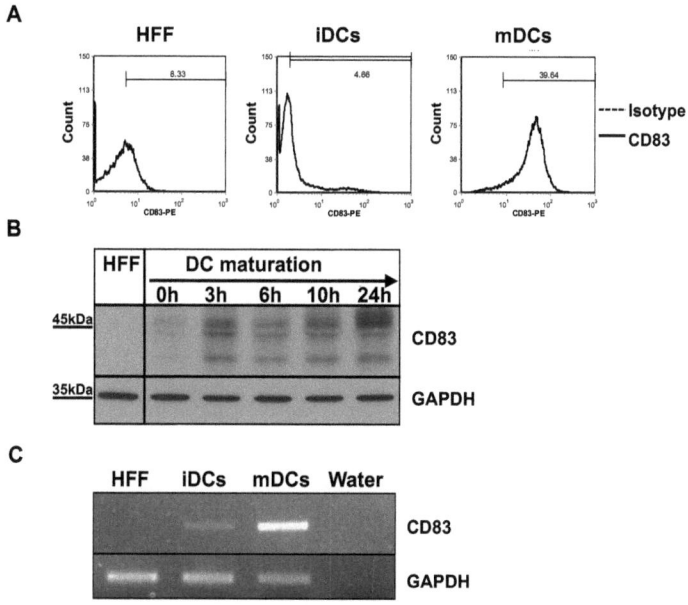

Figure 5.1. CD83 is upregulated during DC maturation and is not expressed by HFF cells.
(A) HFF cells, monocyte derived iDCs, and 20 h LPS-matured monocyte derived mDCs (1*10⁵ each) were harvested and stained with PE-labeled mouse anti-human CD83 mAB or mouse IgG1 isotype-control for flow cytometric analyses. Each quadrangle represents 1 x 10⁴ living cells determined by a gate on forward and side light scatter and PI-staining. In each histogram plot the broken line represents mean fluorescence intensity of the isotype-control stained cells and the solid line represents mean florescence intensity of the anti-CD83 stained cells. (B) For Western blot analyses whole cell lysates from either HFF cells, monocyte derived mDCs or maturing monocyte derived DCs (3, 6, 10 or 24 h after LPS addition) were used. The nitrocellulose membranes were stained with a rat anti-human CD83 and a mouse anti-human GAPDH antibody (loading control). The signal was detected via chemiluminescence. (C) Total RNA was isolated from HFF cells, monocyte derived iDCs and 20 h LPS-matured monocyte derived mDCs and reversely transcribed with an Oligo (dT)$_{12-18}$ primer. Afterwards PCR was performed with an intron spanning primer pair for CD83 and a control primer pair for GAPDH. Additionally, one PCR reaction was prepared with water instead of DNA as a negative control. Data generated in cooperation with Dr. I. Knippert[422].

5.1.2. A six kB region of CD83 intron 2 is specifically acetylated in mDCs

In 2002 the sequence of the human CD83 minimal promoter MP -261 was published by Berchtold and colleagues[167]. However, the MP -261 showed neither cell type- nor maturation status-specific activity. Thus, additional regulatory elements were proposed that regulate the specific gene expression of CD83 in human DCs. Therefore, Ilka Knippertz and colleagues performed chromatin immunoprecipitation-chip (ChIP-chip™) microarray analyses to assess the cell type- and maturation status-specific activation state of the human CD83 gene[422]. This assay is based on the knowledge that reversible acetylation is an important mechanism to regulate gene expression and that hyperacetylated genomic regions suggest high transcriptional activity. Thus, a ChIP-chip[TM] microarray directed against the lysine 9 acetylated histone 3 (H3K9) was performed in collaboration with the German Resource Centre for Genome Research (RZPD, Berlin) and NimbleGen Systems (Reykjavik, Iceland) in order to identify regions of the CD83 gene locus that are specifically acetylated. Therefore, monocyte derived iDCs were matured with a maturation cocktail (MC) containing TNF-α, IL-1β, IL-6 and PGE$_2$ or were left untreated. Next, chromosomal DNA from HFF cells, iDCs and MC-matured DCs was isolated, fragmented and subsequently immunoprecipitated with an antibody specific for the K9 acetylated histone 3. Thereafter, the immunoprecipitated DNA was amplified by a two-step PCR and the enrichment or depletion of the DNA sections in the immunoprecipitated DNA of interest was analyzed by qPCR (data not shown). Hybridization of the DNA fragments on the chip was then performed by the German Resource Centre for Genome Research (RZPD, Berlin) in cooperation with NimbleGen Systems (Reykjavik, Iceland). The fragments of the "Input"-DNA and the "IP"-DNA were labeled with Cy-Dye 5 (Cy5) and Cy-Dye 3 (Cy3), respectively, hybridized and scanned. Afterwards, data were extracted and a preliminary data analysis was performed: The specific enrichment of the hyperacetylated regions ("Input" vs. "IP" DNA) for the CD83 gene locus was aligned to the human genome reference sequence (NCBI Build 36.1). These raw data were further analyzed in cooperation with Stefan Lang (Nikolaus-Fiebiger-Center for Molecular Medicine, Hematopoiesis Unit, Erlangen) using the Signal Map software (*NimbleGen*) and finally interpolated over a space of 500 bp to highlight the differences in hyperacetylation for the different cell types. The analysis revealed that the first 6 kb of the intron 2 of the CD83 gene are specifically hyperacetylated in mDCs at H3K9, whereas there was no acetylation in iDCs and HFF (Fig. 5.2.). Furthermore, no hyperacetylation was detected up to 150 kb upstream and downstream of the CD83 gene (data not shown).In summary, these data indicated a high cell type- and status-specific transcriptional activity within this 6 kb region especially in mDCs providing the basis for further analyses.

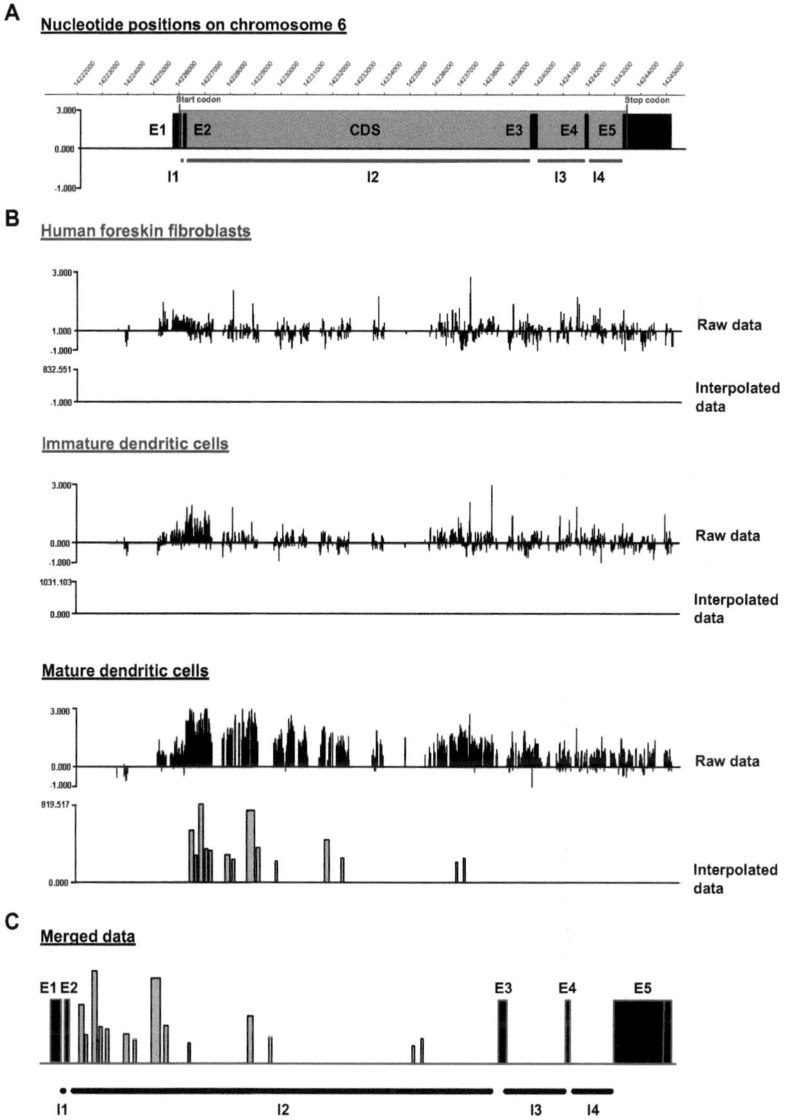

Results

A Nucleotide positions on chromosome 6

B Human foreskin fibroblasts

Immature dendritic cells

Mature dendritic cells

C Merged data

Figure 5.2. Schematic depiction of the CD83 gene locus and the respective acetylation data resulting from the ChIP-chip™ microarray analysis. (A) Schematic depiction of the CD83 genelocus and nucleotide positions on chromosome 6. The coding sequence (CDS) is marked as grey box, exons in black (E1-E5) and introns as bars (I1-I4). **(B)** Data for the CD83 hyperacetylation were provided by NimbleGen Systems and interpolated subsequently over a space of 500 bp using the Signal Map software (NimbleGen). **(C)** Overlay of the interpolated acetylation data for mature DCs aligned with the positions of the CD83 exons (E1-E5) and introns (I1-I4). Data generated in cooperation with Dr. I. Knippertz[422].

5.2. Analyses of the acetylated region of intron 2 within the CD83 gene by luciferase reporter assays

Since the first 6 kb of the CD83 intron 2 were specifically hyperacetylated at H3K9 in mDCs, cell type- and maturation status-specific regulatory elements were assumed to be located within this region. In order to identify these potential regulatory elements, luciferase reporter assays were performed by using the pGL3 based luciferase reporter system (Fig. 5.3.). The pGL3/Basic vector contains a luciferase gene lacking a promoter sequence and was used to assess background transcriptional activity. A pGL3/Basic based vector containing the CMV promoter driving the luciferase gene (pGL3/CMV/luc; kindly obtained from D.M. Nettelbeck, DKFZ & Department of Dermatology; University Hospital Heidelberg) was used as a positive control and for internal normalization. The pGL3/CMV/GFP contains a GFP coding sequence instead of the luciferase gene and was used for the analyses of transfection efficiency. The pGL3/MP-261 vector is based on the pGL3/Basic plasmid, and contains the MP -261 driving the luciferase gene. To assess the enhancing ability of a putative regulatory element (*Put.reg.*) on the MP -261, the DNA sequence of interest was cloned directly upstream of the MP -261 into the pGL3/MP-261 plasmid in sense and antisense orientation, resulting pGL3/Put.reg./MP-261. These reporter plasmids were then used for the transfection of different cell types to perform luciferase reporter assays.

Results

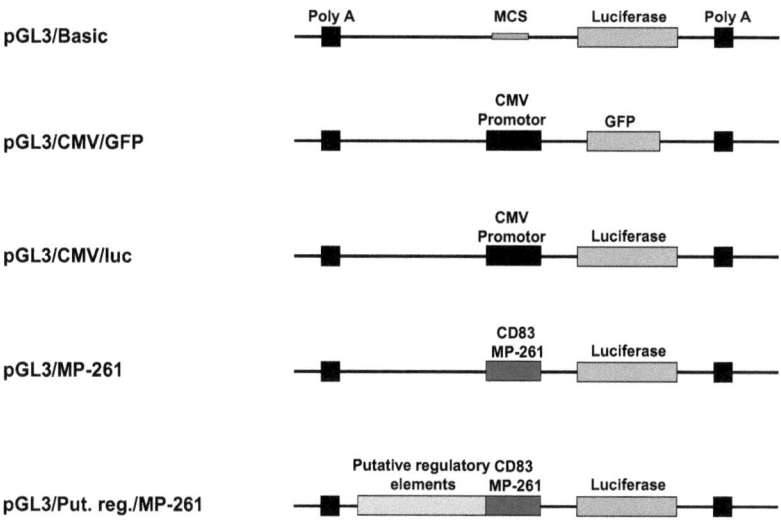

Figure 5.3. Schematic depiction of constructs for luciferase reporter assays. A pGL3 vector backbone only coding for the luciferase gene and lacking any regulatory element (pGL3/Basic) was used for background control. The pGL3 backbone containing the luciferase gene driven by a CMV promoter (pGL3/CMV/luc) was used as a positive control and for internal normalization. The same plasmid coding for GFP instead of luciferase (pGL3/CMV/GFP) was used to determine transfection efficiency. As reference control a pGL3 plasmid coding for the luciferase gene driven by the CD83 minimal promoter MP -261 (pGL3/MP -261) was generated. Finally, plasmids containing a putative regulatory element in both orientations upstream of the MP -261 (pGL3/Put. reg./MP -261) were generated to study its effects on the MP -261.

To narrow down the putative regulatory elements in the first 6 kb of CD83 intron 2, this region was subdivided into three fragments, namely fragment A (1239 bp), B (2359 bp) and C (2220 bp). After PCR amplification, these fragments were subcloned in sense (s) and antisense (as) orientation into the pGL3/MP-261 plasmid directly upstream of the MP -261 (Fig. 5.4.). This resulted in six luciferase reporter constructs, namely: pGL3/Fragment A s/MP-261, /Fragment A as/MP-261, /Fragment B s/MP-261, /Fragment B as/MP-261, /Fragment C s/MP-261 and /Fragment C as/MP-261.

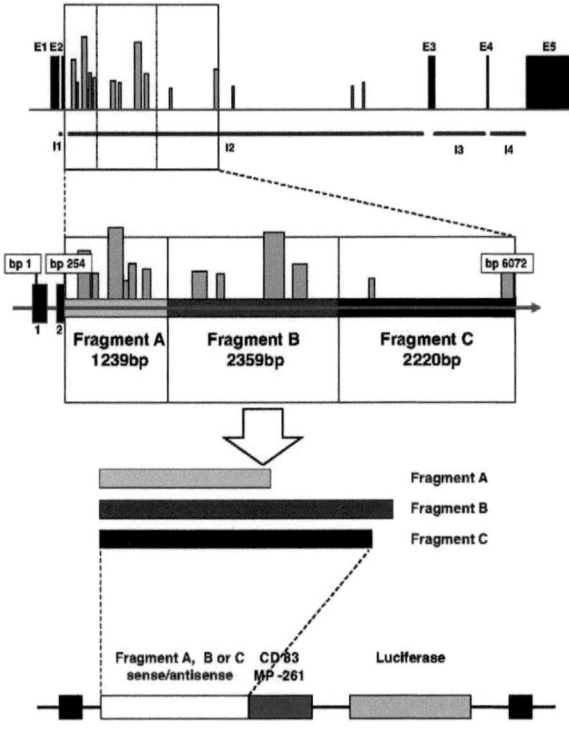

Figure 5.4. Schematic depiction of the subcloning strategy of the hyperacetylated region of CD83 intron 2. Hyperacetylated regions are depicted as grey bars, exons as E1 to E5 and introns are marked as I1 to I5. The boxes indicate the base pair position relatively to the transcription start (bp 1).The first 6 kb of CD83 intron 2 were subdivided into 3 fragments A (1239 bp), B (2359 bp), and C (2220 bp). After PCR amplification each fragment was subcloned in sense and antisense orientation upstream the MP -261 into the pGL3/MP-261 plasmid.

5.2.1. CD83 intron 2 fragment C shows an enhancing activity on MP -261 in the DC-like cell line XS52

In order to investigate, if either fragment A, B or C induce or suppress the activity of the MP -261 in a cell type-specific manner, three different cell lines were transfected with the luciferase reporter constructs described before (see 5.2.). The murine DC-like cell line XS52 served as model for DCs[423], whereas the murine fibroblast cell line NIH3T3 and the human cervix carcinoma cell line HeLa served both as a negative control.

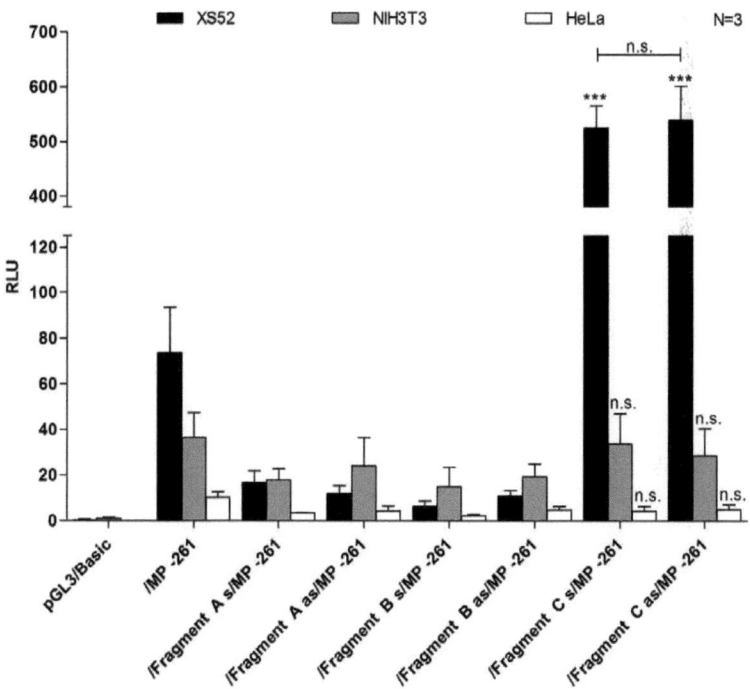

Figure 5.5. Fragment C enhances the MP -261 activity specifically in XS52 cells. XS52, NIH3T3 or HeLa cells were transfected with 2.5 µg of reporter plasmids containing fragment A, B or C of the CD83 intron 2 in sense (s) or antisense (as) orientation upstream of the MP -261 or the pGL3/MP -261 as reference control. The pGL3/Basic and the pGL3/CMV/luc vector were transfected to assess background activity and for internal normalization, respectively. The promoter activity was determined by luciferase reporter assays 48 h after transfection. The resulting RLU were further normalized to the protein concentration of each lysate. The plasmids containing fragments A, B and C were compared to the pGL3/MP -261 vector for each cell line and the resulting p-values were determined via one way ANOVA and Bonferroni's Multiple Comparison *post hoc* test. * $p < 0.05$, ** $p < 0.01$, *** $p < 0.001$ and n.s. not significant ($p > 0.05$). Results represent the means (±SEM) of three independently performed experiments.

Figure 5.5. shows that the background activity of pGL3/Basic in XS52, NIH3T3 and HeLa cells were 0.5 RLU, 1.2 RLU and 0.1 RLU, respectively. The MP -261 displayed transcriptional activity in XS52 (74 RLU), NIH3T3 (36 RLU) and HeLa (10 RLU) cells, which was significantly over background. Fragment A and B did not enhance the MP -261 activity significantly in comparison to the MP -261 alone, regardless of their orientation or cell line. However, fragment C significantly increased the MP -261 activity in XS52 cells in both orientations

Results

(sense: 526 RLU; antisense: 539 RLU), whereas no significant induction of the MP -261 was monitored in NIH3T3 (s: 33.5 RLU; as: 28.4 RLU) or HeLa cells (s: 4.4 RLU; as: 5.0 RLU).

Taken together, fragment C induces the activity of the MP -261 only in XS52 cells. Furthermore, the induction is orientation independent, as there is no significant difference in induction between sense and antisense direction. This hints at a regulatory sequence with the function of a genetic enhancer embedded within fragment C of CD83 intron 2.

5.2.2. Narrowing down the putative enhancer sequence: A 185 bp sequence within fragment C shows enhancer function

In order to narrow down the regulatory sequence embedded within this fragment C, 14 deletion mutants (C1-C14; Fig. 5.6.) were generated and cloned in sense and antisense orientation upstream of the MP -261 into the pGL3/MP-261 luciferase reporter vector. The resulting reporter constructs were transfected into XS52, NIH3T3 and HeLa cell lines. The induction of the MP -261 was assessed 48 h after transfection by luciferase assay in all three cell lines. Figure 5.6. schematically represents the induction of the MP -261 with the various deletion mutants C1-C14 in XS52, NIH3T3 and HeLa cells.

sense and antisense 5'- -3'		XS52	NIH3T3	HeLa
	Fragment C bp 1-2220	+ +	- - -	- - -
	C1 bp 1-1720	+ +	- - -	- - -
	C2 bp 1-1010	+ + +	- - -	- - -
	C3 bp 1-525	+ + +	- - -	- - -
	C4 bp 500-2220	- - -	- - -	- - -
	C5 bp 500-1720	- - -	- - -	- - -
	C6 bp 1000-1720	- - -	- - -	- - -
	C7 bp 100-510	+ + +	- - -	- - -
	C8 bp 1-405	- - -	- - -	- - -
	C9 bp 100-405	- - -	- - -	- - -
	C10 bp 1-300	- - -	- - -	- - -
	C11 bp 100-300	- - -	- - -	- - -
	C12 bp 225-510	+ + +	- - -	- - -
	C13 bp 325-510	+ + +	- - -	- - -
	C14 bp 425-510	+	- - -	- - -

Results

Figure 5.6. Schematic depiction of the deletion mutants C1-C14 of fragment C. The deletion mutants C1-C14 of the CD83 intron 2 fragment C were cloned in sense (s) and antisense (as) orientation into the pGL3/MP-261 reporter plasmid upstream of the MP -261. Induction of the MP -261 activity was determined in XS52, NIH3T3 and HeLa cell lines by luciferase reporter assays. Induction is depicted relatively to the activity of the pGL3/MP -261. - -- no induction of the MP -261, + weak induction of the MP -261, ++ medium induction of the MP -261; +++ strong induction of the MP -261.

Fragment C of CD83 intron 2 showed a specific induction of the MP -261 in XS52, but not in the control cell lines NIH3T3 and HeLa. Deletion of up to 1720 bp of the 3' end of fragment C (C3) did not negatively affect the promoter inducing effect in XS52 cells, whereas deletion of a 500bp fragment at the 5' end of fragment C led to a complete loss of induction (C4). This suggests that a potential regulatory element is embedded within the first 500bp of CD83 intron 2 fragment C. Further deletions were performed in 100 to 125 bp steps. Deletion of up to 225 bp at the 5' end of the remaining 500 bp fragment did not affect the cell type-specific induction of the MP -261 (C13). A further deletion of 100 bp led to a significant loss of luciferase activity (C14). On the other hand a deletion of 120 bp at the 3' end of the remaining 500 bp fragment led to a complete loss of MP -261 induction (C8). Thus, a 185bp region (*185 bp enhancer*) within CD83 intron 2 fragment C (C13) was identified, which induces the MP -261 activity in both orientations specifically in XS52 cells, whereas no induction of the MP -261 was observed in NIH3T3 and HeLa cells.

Figure 5.7. depicts the means of three independent experiments in XS52 cells combining the most important deletion mutants (C1-C4, C7,C8 and C12-C14), though all in figure 5.6. listed deletion mutants were tested in separate experiments (data not shown).

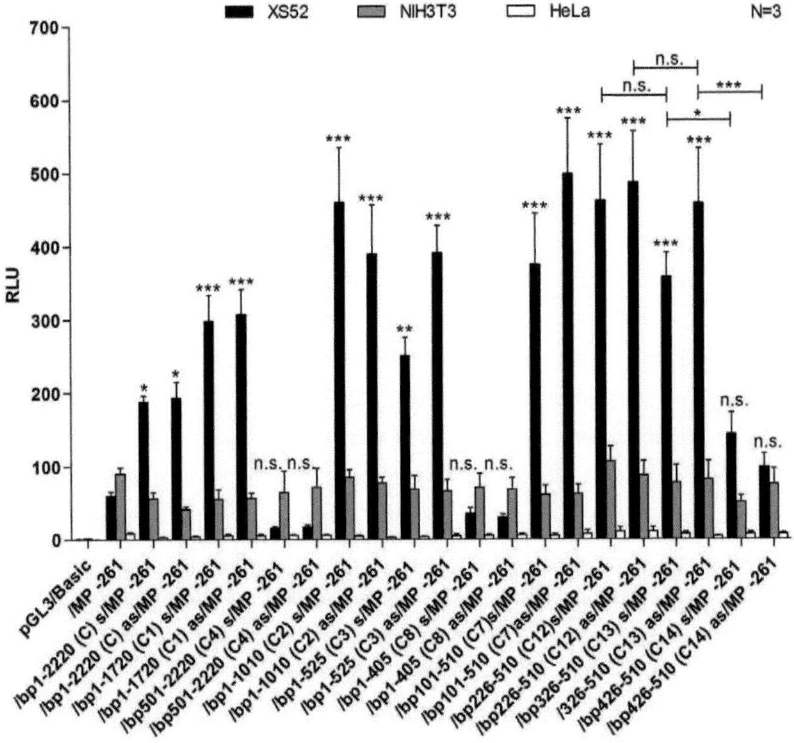

Results

Figure 5.7. The 185 bp long deletion mutant C13 of fragment C enhances the MP -261 activity specifically in XS52 cells. XS52, NIH3T3 or HeLa cells were transfected with 2.5 µg of either pGL3/MP -261 reporter plasmids containing the deletion mutants C1-C14 of the fragment C of the CD83 intron 2 in sense (s) or antisense (as) orientation upstream of the MP - 261 or the pGL3/MP -261 as reference control. The pGL3/Basic and the pGL3/CMV/luc vector were transfected to assess background activity and for internal normalization, respectively. The promoter activity was determined by luciferase reporter assays 48 h after transfection. The resulting RLU were further normalized to the protein concentration of each lysate. The plasmids containing the different fragments were compared to the pGL3/MP -261 vector for each respective cell line and the resulting p-values were determined via one way ANOVA and Bonferroni's Multiple Comparison *post hoc* test. * p < 0.05, ** p < 0.01, *** p < 0.001 and n.s. not significant (p>0.05). Results represent the means (±SEM) of three independently performed experiments.

The background activity of the pGL3/Basic vector amounted to 0.8 RLU in XS52 cells, to 2.0 RLU in NIH3T3 cells and to 0.5 RLU in HeLa cells. The basal activity of the MP -261 added up to 60 RLU, 90.2 RLU and 18.5 RLU in XS52,

NIH3T3 and HeLa cells, respectively. Fragment C induced the MP -261 basal activity specifically in XS52 cells in sense orientation up to 188.0 RLU and in antisense orientation up to 192.9 RLU. Deletion of 1720 bp of the 3' end of fragment C did not affect the induction negatively, but even increased the induction of the MP -261 in sense orientation up to 251.4 RLU and in antisense orientation up to 390.9 RLU (C3; bp 1-525). On the other hand deletion of a 500 bp fragment at the 5' end led to a complete loss of induction (C4; bp 501-2220; 15.5 RLU sense and 17.7 RLU antisense orientation). Further narrowing down was performed in 100 to 125 bp steps. Deletion of 120 bp at the 3' end again led to a complete loss of induction (C8; bp 1-405; 36.0 RLU sense and 30.6 RLU antisense orientation). Deletion of up to 315 bp at the 5' end did not affect significantly the induction of MP -261 (C13; bp 326-510) in both orientations (sense 358.1 RLU; antisense 458.9 RLU). Further deletion of 100 bp led to a significant loss of CD83 promoter induction (C14; bp 426-510) in sense orientation (143.9 RLU) and antisense orientation (98.8 RLU). This remaining induction was not significant when compared to the pGL3/MP -261 control. None of the tested fragments induced significantly the MP -261 activity in NIH3T3 or HeLa cells (tested with ANOVA, but not shown in figure 5.6).

Summing up, deletion mutagenesis of the CD83 intron 2 fragment C results in a 185 bp fragment (*185 bp enhancer*) that induces the MP -261 activity specifically in XS52 cells in sense and antisense orientation up to 7.5 fold, whereas in the control cell lines NIH3T3 and HeLa no induction is observed. This indicates that the enhancing effect might be cell type-specific.

5.2.3. The CD83 intron 2 fragment C and the *185 bp enhancer* induce the MP -261 in human DCs

In order to examine the data obtained from the DC-like cell line XS52 in human primary cells monocyte derived iDCs were electroporated with the pGL3/Mp-261 reporter plasmids, either containing fragment C or the *185bp enhancer* in sense or antisense orientation upstream of the MP -261 or lacking the additional sequence. After electroporation, cells were divided into two equal fractions. Three hours after electroporation, one of the cell fractions was matured for 20 h with LPS, whereas the other fraction was left untreated. The next day, cells were lysed and the induction of the MP -261 by either fragment C or the *185 bp enhancer* was assessed by a luciferase reporter assay.

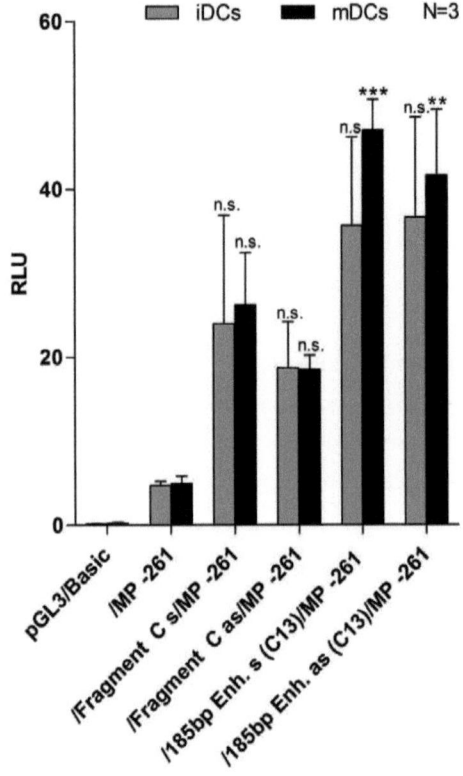

Figure 5.8. Fragment C and the *185 bp enhancer* induce the MP -261 activity specifically in mDCs. Monocyte derived iDCs were electroporated with 4 µg of either pGL3/Basic, pGL3/MP-261, pGL3/Fragment C/MP-261 s/as or pGL3/185bpEnh.(C13)/MP-261 s/as luciferase reporter plasmid. Afterwards, cells were transferred to a 12-well tissue culture plate and split into two equal fractions. One of the fractions was matured 3 h after electroporation with LPS to a final concentration of 0.1 ng/ml and the other fraction was replenished with cell culture medium without LPS. The pGL3/Basic and the pGL3/CMV/luc vector were electroporated to assess background activity and for internal normalization, respectively. The promoter activity was determined by luciferase reporter assays 20 h after LPS addition and the resulting RLU were normalized to the protein concentration of each lysate. The plasmids containing the different fragments were compared to the pGL3/MP -261 vector and the resulting p-values were determined via one way ANOVA and Bonferroni's Multiple Comparison *post hoc* test. * p < 0.05, ** p < 0.01, *** p < 0.001 and n.s. not significant (p>0.05). Results represent the means (±SEM) of three independently performed experiments.

As shown in figure 5.8., the background activity of the pGL3/Basic vector in iDCs and mDCs amounted to 0.1 RLU and 0.2 RLU, respectively. The basal activity of the MP -261 was well over background and did not vary significantly between immature (4.7 RLU) and mDCs (5.0 RLU). Albeit fragment C induced the MP -261 activity in iDCs in both orientations (sense: 24.0 RLU; antisense: 18.7 RLU), the induction was not statistically significant. In a similar fashion, fragment C induced the MP -261 in sense orientation (26.2 RLU) and in antisense orientation in mDCs (18.5 RLU). However, this induction was also not statistically significant. Interestingly, the difference between iDCs and mDCs in the induction of the MP -261 was much more evident with plasmids containing the *185 bp enhancer*. Although the enhancer induced the MP -261 to some extent in iDCs (sense: 35.6 RLU; antisense: 36.6 RLU), the induction was not statistically significant. In strong contrast, the induction of the MP -261 by the *185 bp enhancer* was highly significant in mDCs in both sense (47.0 RLU; p<0.001) and antisense (41.6 RLU; p<0.01) orientation.

Taken together, the CD83 intron 2 fragment C and the 185 bp enhancer significantly induce the MP -261 only in monocyte derived mDCs. Although both elements also induce the MP -261 to some extent in iDCs, the induction is not statistically significant. This induction may be attributed to a partial maturation of the iDCs (surface markers checked by FACS; data not shown) caused by the physical stress of the electroporation procedure.

5.2.4. CD83 intron 2 fragment C from and the *185 bp enhancer* do not induce the MP -261 in B and T cell lines

Specific subtypes of B and T cells also express CD83 in a status-specific manner[123;424]. Thus, next it was elucidated, whether the 185bp enhancer induces MP -261 activity also in a B cell line (Raji) or a T cell line (Jurkat) in the same way as in mDCs. FACS analyses revealed that both cell lines moderately express CD83 (data not shown). Therefore, Raji B cells and Jurkat T cells were electroporated with the pGL3/Mp-261 reporter plasmids, either containing fragment A, B, C or the 185bp enhancer in sense or antisense orientation upstream of the MP -261 or lacking the additional fragments. Two days after electroporation cells were lysed in reporter lysis buffer and the induction of the MP -261 was assessed by luciferase assays.

Results

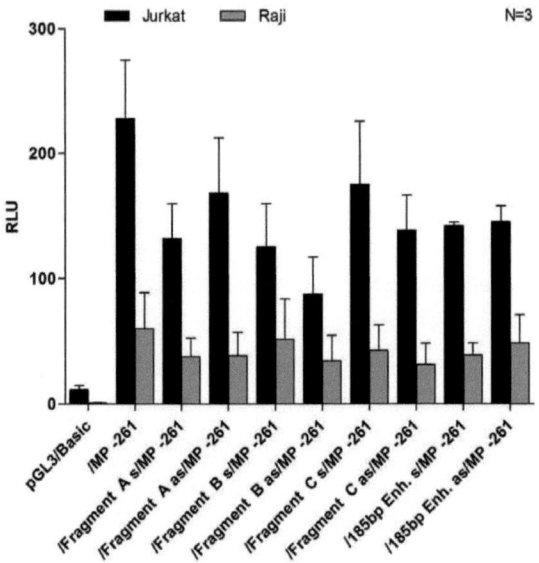

Figure 5.9. Neither fragments A, B, C nor the *185 bp enhancer* induce the MP -261 in Raji and Jurkat cells. Raji or Jurkat cells were electroporated with 20 µg of reporter plasmids containing fragment A, B, C or the *185 bp enhancer* in sense (s) or antisense (as) orientation upstream of the MP -261 or the pGL3/MP -261 as reference control. The pGL3/Basic and the pGL3/CMV/luc vector were electroporated to assess background activity and for internal normalization, respectively. The promoter activity was determined 48 h after electroporation by luciferase reporter assays and the resulting RLU were further normalized to the protein concentration of each lysate. The plasmids containing the different fragments were compared to the pGL3/MP -261 vector for each respective cell line and the resulting p-values were determined via one way ANOVA and Bonferroni's Multiple Comparison *post hoc* test. Results represent the means (±SEM) of three independently performed experiments.

The background activity of the pGL3/Basic vector in Raji and Jurkat cells amounted to 11.3 RLU and 0.8 RLU, respectively. The CD83 minimal promoter (MP -261) alone displayed a luciferase activity of 228.2 RLU in Jurkat and 59.8 RLU in Raji cells. No induction of the MP -261, neither by fragment A, B, C nor the 185bp enhancer was observed for both cell lines, with values ranging from 150 to 180 RLU in Jurkat and values ranging from 40 to 50 RLU in Raji cells. This indicates that - although Raji and Jurkat cells both express CD83 - the enhancer is not active in these cell lines and is thus specific in mDCs.

5.3. Biocomputational analyses and modelling of the human CD83 promoter

As shown above, the 185 bp enhancer induces the MP -261 activity in a mDC-specific manner. In order to investigate the underlying molecular mechanisms, first biocomputational analyses of the ~500 bp long area upstream of the CD83 transcription start, including the MP -261 and the 185 bp enhancer (bp 326 to 510 of CD83 intron 2 fragment C) were performed in cooperation with Dr. Thomas Werner (Munich). A depiction of the results obtained from the biocomputational analyses is shown in figure 5.10.

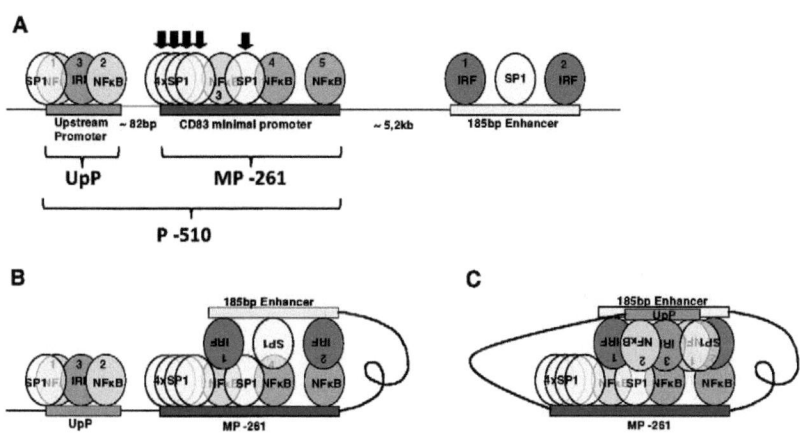

Figure 5.10. Biocomputational model depicting the interaction of the CD83 upstream promoter, the MP -261 and the 185 bp enhancer. (A) Depiction of the three regulatory elements with the predicted transcription factor binding sites regulating CD83 transcription: (i) The upstream promoter (UpP) containing NFκB-sites 1 and 2, IRF-site 3 and one SP1-site, (ii) the 261 bp long CD83 minimal promoter (MP -261) containing NFκB-sites 3, 4 and 5 and four SP1-sites and (iii) the 185 bp enhancer containing IRF-sites 1 and 2 as well as one SP1-site. The UpP and the MP -261 including the 82 bp naturally occurring spacer sequence are termed 510 bp promoter (P -510). Previously confirmed and published transcription factor binding sites are marked with an arrow ⬇. (B) Proposed interaction of the 185 bp enhancer with the MP -261 mediated by the interaction of IRF-sites 1 and 2 with NFκB-sites 3 and 5, respectively. (C) Ternary complex formation with the UpP as third interaction partner. The interaction between the UpP and the MP -261 is mediated by IRF-site 3 and NFκB-site 4.

The predicted NFκB-sites were numbered from 1 (upstream) to 5 (downstream) according to their order in the analyzed genomic region. The predicted IRF-sites in the *185 bp enhancer* were numbered 1 (upstream) and 2 (downstream). Number 3 was assigned to the predicted IRF-site in the UpP. The computational sequence analyses of the CD83 *185 bp enhancer* and the MP -261 confirmed the already published NFκB-site 4 and four SP1-sites within the MP -261[167] (Fig. 5.10. A). In addition, the biocomputational analyses predicted two NFκB-sites (NFκB-sites 3 and 5) and one further SP1-site for the MP -261. For the *185 bp enhancer* two IRF-sites (IRF-sites 1 and 2) and one SP1-site were predicted. As depicted in figure 5.10. (B), the model proposes an interaction between the MP -261 and the *185 bp enhancer,* mediated by the interaction of IRF-site 1 with NFκB-site 3, as well as the interaction of IRF-site 2 with NFκB-site 4. The SP1-sites are supposed to function as stabilizing elements. Thus, NFκB-site 4, albeit being described in the literature as essential, is not included into the formation of the 185 bp enhancer/MP -261 complex. Hence, the region upstream of the MP -261 was analyzed for additional TFBs complementing NFκB-site 4. Thereby, two NFκB-sites (NFκB-sites 1 and 2) as well as one IRF-site (IRF-site 3) were predicted for the putative upstream promoter (UpP; Fig. 5.9. A). At this point the predicted UpP is thought to loop in to fill in the gap (Fig. 5.10. C). Importantly, the distance of 83 bp between UpP and MP -261 perfectly allows for looping of the DNA in the described manner. Thereby, IRF-site 3 and NFκB-sites 1 and 2 in the UpP are presumed to interact with the NFκB-site 4 in the MP -261 and the IRF-sites 1 and 2 in the *185 bp enhancer*, respectively. In this context, the NFκB-sites 1, 2 and 3 as well as the IRF-site 3 were considered as weak binding sites, attracting the respective transcription factors probably through cooperative binding.

Taken together, the bioinformatical model predicts the interaction of NFκB-sites 3 and 5 in the MP -261 with IRF-sites 1 and 2 in the *185 bp enhancer*. Furthermore, IRF-site 3 as well as NFκB-sites 1 and 2 in the UpP are assumed to interact with NFκB-site 4 in the MP -261 and IRF-sites 1 and 2 in the enhancer, respectively, thus forming three IRF-NFκB-IRF modules. Thus, a ternary complex involving all predicted transcription factor binding sites is formed, providing a potent platform to initiate CD83 transcription.

5.4. Functional characterization of the predicted UpP and the spacer sequence S1 within the pGL3/MP -261 reporter plasmids

To provide functional evidence for the formation of a ternary complex of UpP, MP -261 and *185 bp enhancer*, reporter constructs were generated containing the UpP in genomic configuration 83 bp upstream of the MP -261 (termed P - 510). Furthermore, a 500 bp spacer sequence was cloned between the MP - 261 or the P -510 and the *185 bp enhancer* (Fig. 5.11.) for the following reason: The spacer sequence minimizes intermolecular plasmid-plasmid interactions, which may have compensated the lack of the UpP in previous experiments, as the missing IRF-site could have been provided by the *185 bp enhancer* of a neighboring plasmid. The spacer sequence allows the bending of the DNA, facilitating thereby the *cis* enhancer-promoter interaction. Furthermore, in an adenoviral setting the spacer sequence allows an intramolecular bending of the DNA to facilitate the enhancer-promoter interaction.

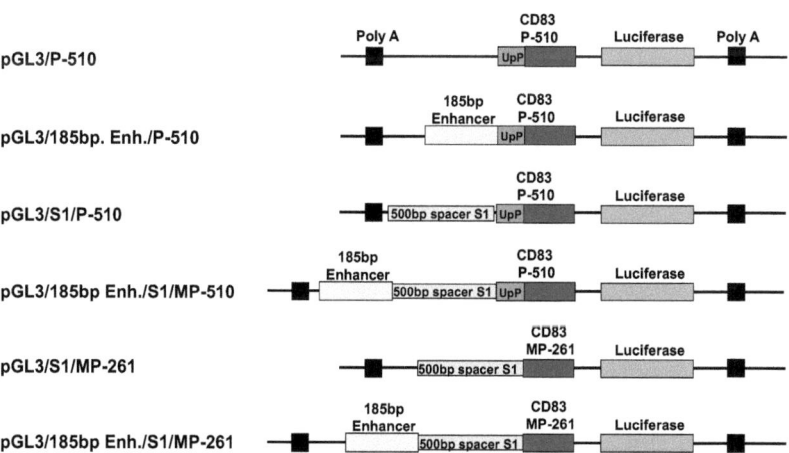

Figure 5.11. Schematic depiction of reporter constructs containing the UpP and the spacer sequence S1. The P -510 as well as the UpP were cloned upstream of the luciferase gene into the pGL3/Basic plasmid, thus generating pGL3/P-510 and pGL3/UpP, respectively. The UpP comprised in the P -510 is highlighted in purple. Furthermore, a 500 bp long spacer sequence (S1) was inserted between MP -261 and P -510, generating the plasmids pGL3/185 bp Enh./S1/MP-261 and pGL3/185 bp Enh./S1/MP -261, respectively. The spacer sequence was also introduced into the plasmids pGL3/MP-261 and pGL3/P-510, resulting in the plasmids pGL3/S1/MP-261 and pGL3/S1/P-510.

Based on the previous reporter constructs (Fig. 5.3.), plasmids comprising the P -510 instead of the MP -261 were generated either containing the *185 bp enhancer* or not. Therefore, the MP -261 was enzymatically removed from the pGL3/185bp Enh./MP-261 plasmids and substituted by the P -510. P -510 contains both the predicted UpP and the MP -261 in a genomic configuration (see Fig. 5.10.). Furthermore, a 500 bp long spacer sequence (S1) was inserted into the reporter constructs containing the MP -261 or the P -510 with or without the *185 bp enhancer*. The spacer sequence S1 was generated by PCR amplification from a non coding region located 3 kb upstream of the CD83 transcription start site and cloned between the MP -261 or P -510 and the *185 bp enhancer*.

5.4.1. Neither the spacer sequence S1, nor the UpP do significantly affect the induction of the CD83 promoter in cell lines

To exclude any regulatory side effect of the newly introduced spacer sequence S1 on either the MP -261 or the P -510, the corresponding constructs were compared to those without spacer in a luciferase reporter assay. Therefore, XS52, NIH3T3 and HeLa cells were transfected with the luciferase reporter plasmids described in figure 5.11. as well as with the constructs lacking the spacer S1. Their activity was investigated using a luciferase assay 48 h after transfection.

Results

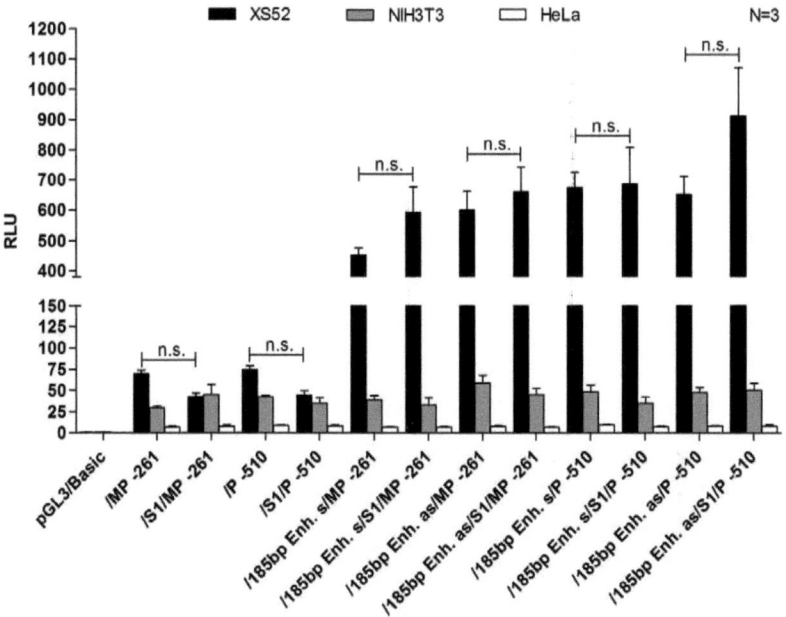

Figure 5.12. The spacer sequence S1 does not significantly affect the induction of the MP -261 and the P -510 in XS52 cells. XS52, NIH3T3 and HeLa cells were transfected with 2.5 µg of either pGL3/MP -261 or pGL3/P -510 with or without the *185 bp enhancer* in sense (s) or antisense (as) orientation either containing the spacer sequence S1 or not. As controls the pGL3/Basic and the pGL3/CMV/luc vector were transfected to assess background activity and for internal normalization, respectively. The promoter activity was determined by luciferase reporter assays 48 h after transfection. The resulting RLU were further normalized to the protein concentration of each lysate. The plasmids containing the spacer sequence S1 were compared to the corresponding plasmids without the spacer sequence for each respective cell line. The resulting p-values were determined via one way ANOVA and Bonferroni's Multiple Comparison *post hoc* test. * $p < 0.05$, ** $p < 0.01$, *** $p < 0.001$ and n.s. not significant (p>0.05). Results represent the means (±SEM) of three independently performed experiments.

The spacer sequence S1 showed no statistically significant effect on the induction of the MP -261 or the P -510 in XS52 cells. The basal activity of the MP -261 construct without spacer resulted in 70.0 RLU in comparison to 42.7 RLU of the construct with the spacer. Comparably, the P -510 construct without spacer resulted in 74.3 RLU compared to 44.5 RLU of the construct with spacer. Similar results were achieved for the MP -261 and P -510 plasmids containing the *185bp enhancer*. The activity of the MP -261 constructs lacking the spacer sequence S1 (453.4 RLU with enhancer in sense and 599.9 RLU in antisense orientation) did not significantly differ from the corresponding constructs containing the spacer (592.2 RLU in sense and 660.7 RLU in antisense orientation). Likewise, the activity of the P-510 constructs was not significantly influenced by the spacer S1, resulting in 674.4 RLU (*185 bp enhancer* sense) and 650.6 RLU (*185 bp enhancer* antisense) for the constructs without spacer compared to 686.5 RLU (s) and 910.5 RLU (as) for the constructs with spacer.

Interestingly, several effects were observed in XS52 cells for the MP -261 plasmids in comparison to plasmids containing the P -510: The basal activity of the MP -261 and P -510 lacking the *185bp enhancer* did not differ significantly, regardless of the spacer sequence S1. Plasmids containing the *185bp enhancer*, showed an overall trend for a stronger, albeit not statistically significant, induction of the P -510 than of the MP -261, indicating an involvement of the presumed UpP in the formation of the predicted ternary complex regulating CD83 transcription. For the NIH3T3 and the HeLa control cells no significant differences or trends in the induction of the MP -261 or the P -510 were observed, regardless of the inclusion of the spacer sequence S1 or the 185 bp enhancer.

Taken together, the addition of the spacer sequence S1 into the luciferase reporter constructs does neither significantly affect the basal activity of the MP -261 and the P -510 nor their induction by the *185 bp enhancer* in XS52 cells. On the other hand, the inclusion of the UpP into the reporter constructs containing the *185 bp enhancer* shows a trend towards a higher induction of the MP -261 in XS52 cells, indicating an involvement of the UpP in the transcriptional regulation of CD83. In control cells NIH3T3 and HeLa, neither the inclusion of the spacer S1 nor the UpP significantly changes the basal activity of the MP -261.

5.4.2. The spacer sequence S1 does not significantly affect the induction of the CD83 promoter in mDCs.

In order to monitor the influence of the spacer sequence S1 on the induction of MP -261 and the P -510 in monocyte derived DCs, iDCs were electroporated with the MP -261 and P -510 constructs including the spacer sequence S1 (see Fig. 5.11.) and the corresponding constructs lacking S1 Three hours after

Results

electroporation cells were matured with LPS for 20 h and the induction of the MP -261 and the P -510 was assessed by luciferase assay.

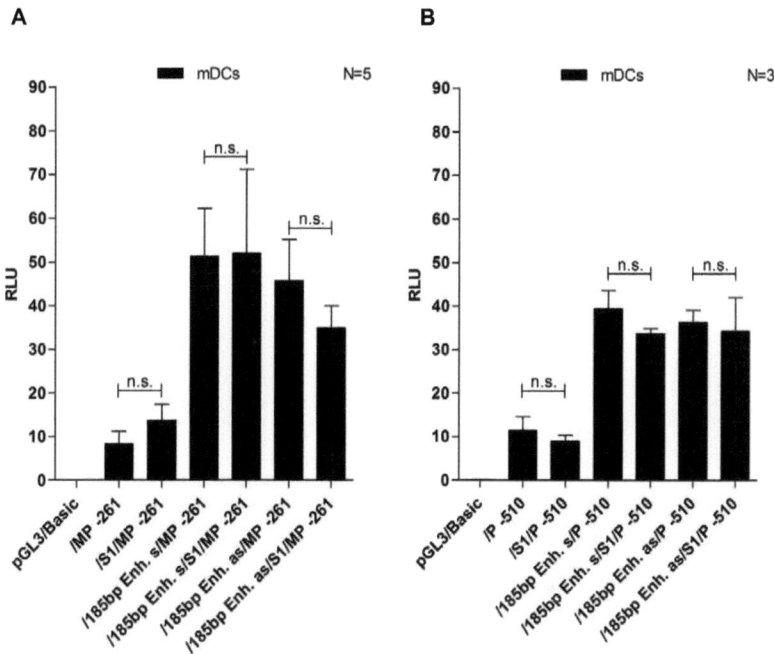

Figure 5.13. The spacer sequence S1 does not significantly affect the induction of the MP -261 in mDCs. (A) Monocyte derived iDCs were electroporated with 4 µg of either pGL3/MP -261with or without the 185 bp enhancer in sense (s) or antisense (as) orientation upstream of the MP -261 or with their respective counterparts comprising the spacer sequence S1. Cells were matured 3 h after electroporation with LPS to a final concentration of 0.1 ng/ml. As controls the pGL3/Basic and the pGL3/CMV/luc vector were electroporated to assess background activity and for internal normalization, respectively. The promoter activity was determined by luciferase reporter assays 20 h after LPS addition and the resulting RLU were further normalized to the protein concentration of each lysate. The plasmids containing the spacer sequence S1 were compared to the corresponding plasmids lacking the spacer sequence. The resulting p-values were determined via one way ANOVA and Bonferroni's Multiple Comparison post hoc test. * p < 0.05, ** p < 0.01, *** p < 0.001 and n.s. not significant (p>0.05). Results represent the means (±SEM) of five independently performed experiments. (B) The spacer sequence S1 does not significantly affect the induction of the P -510 in mDCs. Cells were treated and electroporated in the same way as in (A), but instead of plasmids bearing the MP -261, plasmids comprising the P -510 were used. The resulting p-values were determined via one way ANOVA and Bonferroni's Multiple Comparison post hoc test. * p < 0.05, ** p < 0.01, *** p < 0.001 and n.s. not significant (p>0.05). Results represent the means (±SEM) of three independently performed experiments.

As displayed in figure 5.13. (A) the spacer sequence S1 had no significant influence on the basal activity and the induction of the MP -261 in mDCs. The basal activity of the MP -261 without spacer resulted in 8.3 RLU and was not significantly altered by the introduction of the spacer S1 (13.6 RLU). Likewise, no significant difference in the induction of the MP -261 by the *185 bp enhancer* in presence or in absence of the spacer was observed. The induction by the *185 bp enhancer* in sense orientation without spacer resulted in 51.3 RLU, whereas the induction with the spacer sequence resulted in 52.1 RLU. With the *185 bp enhancer* in antisense orientation, the induction of the MP -261 lacking the spacer resulted in 45.7 RLU, whereas with the spacer it resulted in 35 RLU. Figure 5.13. (B) shows that the results obtained with the P -510 constructs were comparable to those achieved with the MP -261 constructs. The basal activity of the P -510 without spacer S1 resulted in 11.4 RLU, whereas the activity in presence of spacer resulted in 9.0 RLU. Furthermore, the induction of the P -510 by the *185 bp enhancer* in sense orientation without spacer sequence resulted in 39.4 RLU and with spacer in 33.7 RLU. Likewise, the induction of the P -510 by the enhancer in antisense orientation without spacer resulted in 36.3 RLU and with spacer in 34.3 RLU.

In summary, neither the basal activity nor the induction of the MP -261 and the P -510 by the *185 bp enhancer* is altered by the addition of the spacer sequence S1 in a statistical significant manner in mDCs.

5.5. Adenoviral transduction confirms the cell type- and status-specific interaction of UpP, MP -261 and *185 bp enhancer*

As described previously (see 5.2.3.), the physical stress of electroporation partially matured iDCs and led to high amount of cell death. Furthermore, the electroporation of small plasmids entails intermolecular (*trans*) interaction of e.g. the *185 bp enhancer* of one plasmid with the MP -261 of another plasmid rather than facilitating the formation of the ternary complex by intramolecular (*cis*) looping. In this context the adenoviral transduction of DCs with reporter constructs has several advantages over electroporation. First, it has been shown that adenoviral transduction does not interfere with DC-functions. This allows the faithful examination of the promoter activity in DCs in a maturation

Results

status-dependent manner. Secondly, the large adenoviral vectors favor the intramolecular interaction of the promoter and enhancer elements over the intermolecular interactions of small plasmids, thus allowing an examination in a more native conformation. Hence, for steric reasons, all adenoviral vectors included the spacer sequence S1 (see 5.4.).

5.5.1. The proposed ternary complex of UpP, MP -261 and the *185 bp enhancer* forms in mDCs

To generate the plasmids containing the recombinant serotype 5 adenoviral genome the pAdEasy1-system was used. Resulting plasmids were transfected into 293 cells for virus assembly and amplification. Monocyte derived iDCs were transduced with adenoviral vectors containing a luciferase gene under the control of the MP -261 (Ad261/S1) or the P -510 (Ad510/S1) either with or without the *185 bp enhancer* (Fig. 5.14.). To exclude a promoter activity of the enhancer, several control virus constructs were generated. Cells were transduced with AdBasic/S1+Es and AdBasic/S1+Eas, lacking the promoter, but containing the *185 bp enhancer* either in sense or antisense orientation upstream of the luciferase gene. Furthermore, the GFP encoding adenovirus Ad5TL was transduced to assess transduction efficiency. Moreover, cells were transduced with Ad5luc1 expressing a luciferase gene under the control of the CMV promoter as positive control and for internal normalization. Three hours after transduction, cells were either matured with LPS for 20 h or were left untreated and the induction of MP -261 and P -510 was assessed by luciferase assays.

Results

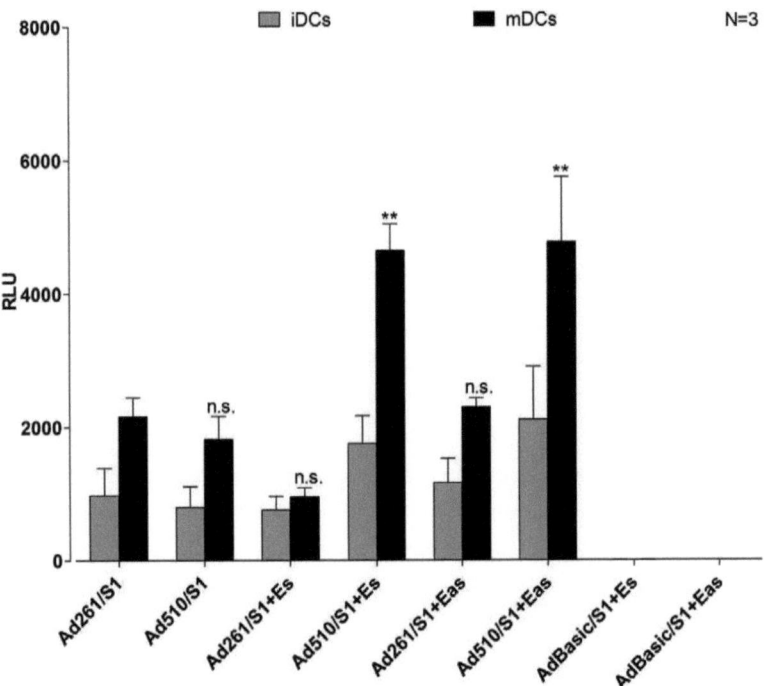

Figure 5.14. The ternary complex of UpP, MP -261 and *185 bp enhancer* shows a specific transcriptional induction in mDCs. Monocyte derived iDCs were transduced with recombinant adenovirus as indicated at 500 $TCID_{50}$/cell. Cells were either matured 3 h after electroporation with LPS to a final concentration of 0.1 ng/ml mDCs) or replenished with cell culture medium without LPS (iDCs). The promoter activity was determined by luciferase reporter assays 20 h after LPS addition. As controls the Ad5luc1 was transduced for internal normalization. The resulting RLU were further normalized to the protein concentration of each lysate. All transduction setups were compared to Ad261/S1 for iDCs and mDC, respectively. The resulting p-values were determined via one way ANOVA and Bonferroni's Multiple Comparison *post hoc* test. * $p < 0.05$, ** $p < 0.01$, *** $p < 0.001$ and n.s. not significant (p>0.05). Results represent the means (±SEM) of three independently performed experiments.

The basal activity of MP -261 and P -510 did not vary significantly in iDCs (982.8 RLU and 801.9 RLU, respectively). The *185 bp enhancer* did neither induce the basal activity of MP -261 nor of P -510 significantly, regardless of its orientation. However, the combination of P -510 and enhancer resulted in a slightly increased RLU (1754.2 RLU sense, 2110.3 RLU antisense) as compared to the MP -261 (763.7 RLU sense 1166.3 RLU antisense). Likewise, in mDCs the basal activity of MP -261 and P -510 did not significantly differ (2157.2 RLU and 1820.8 RLU, respectively), but is clearly higher as opposed to iDCs. Interestingly, the MP -261 was not significantly induced by the *185 bp enhancer*, neither in sense (958.2 RLU) nor in antisense (2291.4 RLU) orientation. It is an important observation that only the P -510 was significantly induced by the enhancer in sense (4647.5 RLU) and antisense (4776.7 RLU) orientation. Accordingly, the presence of the UpP enabled the induction of the MP -261 by the *185 bp enhancer*. Thus, only the combination of all three proposed regulatory elements, namely UpP, MP -261 and *185 bp enhancer* induced the highest transcriptional activity in mDCs. Transduction with AdBasic/S1+Es and AdBasic/S1+Eas revealed that the *185 bp enhancer* alone did not unspecifically induce luciferase expression in iDCs (s: 2.0 RLU; as: 2.1 RLU) or in mDCs (s: 0.3 RLU; as: 0.6 RLU).

Taken together, neither MP -261 nor P -510 are significantly induced by the enhancer in iDCs. On the other hand, in mDCs only the P -510 is significantly induced by the *185 bp enhancer*, whereas the MP -261 is not. This underlines the importance of the UpP for the formation of the predicted ternary complex in a maturation status-dependent manner.

5.5.2. The proposed ternary complex does not form in Raji, Jurkat and JCAM cell lines

It has been shown that CD83 is not only expressed on mDCs, but also on activated B cells and subtypes of T cells[123;424]. As described previously (see chapter 5.2.4.), the *185 bp enhancer* did not induce the MP -261 activity in Raji B cells and Jurkat T cells. Given that the UpP plays a pivotal role for the induction of the MP -261 by the *185 bp enhancer* in DCs, it was analyzed whether the ternary complex also induces transcription in the Raji B cells and the Jurkat- and JCAM-T cells. Therefore, cells were transduced with the same adenoviral constructs described previously for DCs (see chapter 5.5.1.). Moreover, cells were either stimulated with LPS (Raji) and PGE_2/TNF-α (Jurkat and JCAM) after transduction or left untreated. The transduction efficiency, as well as the CD83 surface expression was assessed by FACS (data not shown).

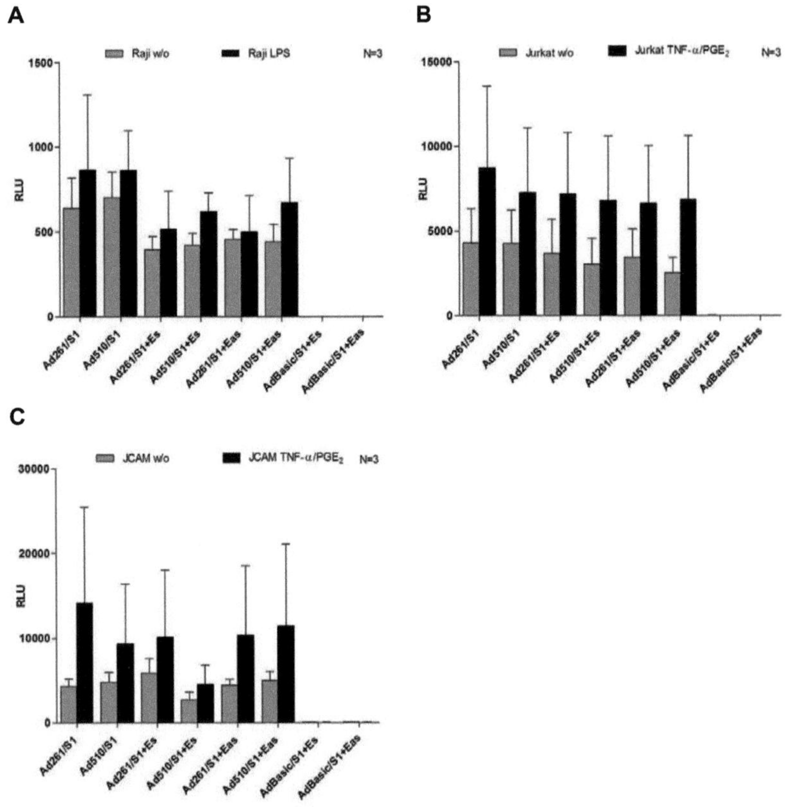

Results

Figure 5.15. The ternary complex of UpP, MP -261 and *185 bp enhancer* shows no specific transcriptional induction in Raji, Jurkat and JCAM cells. (A)-(C) Raji, Jurkat and JCAM cells were transduced with recombinant adenovirus as indicated at 500 TCID$_{50}$/cell for Jurkat and JCAM and 50 TCID$_{50}$/cell for Raji cells. Raji cells were either stimulated 3 h after transduction with LPS to a final concentration of 0.1 ng/ml or left untreated. Jurkat and JCAM cells were either stimulated 3 h after transduction with PGE$_2$ and TNF-α to a final concentration of 1 µg/ml and 10ng/ml, respectively, or left untreated. The promoter activity was determined by luciferase reporter assays 20 h after stimulation. As additional controls Ad5luc1 was transduced for internal normalization. The resulting RLU were further normalized to the protein concentration of each lysate. All transduction setups were compared to Ad261/S1 for each cell line and stimulation, respectively. The resulting p-values were determined via one way ANOVA and Bonferroni's Multiple Comparison *post hoc* test. Results represent the means (±SEM) of three independently performed experiments.

In unstimulated Raji cells (Fig. 5.15. A) both MP -261 and P -510 displayed a similar basal activity (640.6 RLU and 703.5 RLU, respectively), which was weakly induced by LPS stimulation (865.8 RLU and 863.5 RLU, respectively).

Results

The *185 bp enhancer* did not significantly induce MP -261 or P -510, regardless of LPS stimulation. Transduction with AdBasic/S1+Es and AdBasic/S1+Eas revealed that the *185 bp enhancer* alone did not promote luciferase expression in unstimulated (S: 0.5 RLU; as: 0.4 RLU) or LPS-stimulated Raji cells (S: 0.3 RLU; as: 0.3 RLU).

Similar results were obtained for the Jurkat T cells (Fig. 5.15. B). The basal activity of both MP -261 and P -510 (4295.1 RLU and 4266.3 RLU, respectively) did not differ significantly in unstimulated Jurkat cells and both promoters were not significantly induced by the enhancer. Interestingly, the basal activity of both MP -261 and P -510 was strongly increased by the stimulation with TNF-α/PGE$_2$ to 8735.5 RLU and 7263.0 RLU, respectively. However, even in stimulated Jurkat cells neither MP -261 nor P -510 were significantly induced by the *185 bp enhancer*, regardless of its orientation. AdBasic/S1+Es and AdBasic/S1+Eas revealed that the *185 bp enhancer* had no promoter activity in unstimulated (s: 11.5 RLU; as: 5.7 RLU) or LPS-stimulated Jurkat cells (s: 8.0 RLU; as: 3.9 RLU).

As shown in figure 5.15. (C), JCAM cells displayed similar results to Jurkat cells. In unstimulated JCAM cells, both MP -261 and P -510 displayed a similar basal activity (4273.1 RLU and 4705.1 RLU, respectively), which was not significantly induced by the enhancer. Again, the basal activity of both MP -261 and te P -510 was induced by the stimulation with TNF-α/PGE$_2$ to 14106.0 RLU and 9330.3 RLU, respectively. Moreover, also stimulation of JCAM cells did not result in the induction of MP -261 or P -510 by the *185 bp enhancer*, regardless of its orientation. Transduction of AdBasic/S1+Es and the AdBasic/S1+Eas confirmed that the *185 bp enhancer* had no promoter activity in unstimulated (s: 16.3 RLU; as: 11.8 RLU) or TNF-α/PGE$_2$-stimulated JCAM cells (s: 16.1 RLU; as: 12.5 RLU).

In summary, the basal transcriptional activity of the MP -261 is not induced by the UpP and the 185 bp enhancer in Raji, Jurkat and JCAM cells. This underlines the cell type specificity of the ternary complex of UpP, MP -261 and *185 bp enhancer*. However, the basal activity of the MP -261 and thus also the CD83 expression can be upregulated by stimulation with LPS or TNF-α/PGE$_2$ (analyzed by FACS; data not shown) in all three cell types, regardless of the presence of the UpP and the 185 bp enhancer due to the upregulation of NFκB.

5.6. Analyses of the predicted transcription factor binding sites by electrophoretic mobility shift assays (EMSA)

5.6.1. Bandshift analyses of the predicted NFκB- and IRF-sites

The biocomputational analyses predicted a ternary NFκB-IRF transcriptional module encompassing five NFκB- and three IRF-transcription factor binding sites. Furthermore, transduction experiments with adenoviral luciferase reporter vectors confirmed the cell type- and maturation status-specific formation of the predicted ternary complex. To elucidate the molecular mechanism underlying the cooperation of UpP, MP -261 and *185 bp enhancer*, the binding of nuclear proteins to the biocomputationally predicted TFBSs was analyzed by EMSAs. Therefore, oligonucleotides coding for the predicted NFκB-sites 1-5 (Fig. 5.16 A) and the predicted IRF-sites 1-3 (Fig. 5.16 B) were radioactively ^{32}P-end-labeled and incubated with nuclear extracts derived from iDCs, mDCs or HFF cells and loaded onto a non denaturing polyacrylamide gel. For the preparation of the nuclear extracts, monocyte derived iDCs were either matured for 20 h with LPS or were left untreated. The next day, iDCs, mDCs and HFF cells were harvested and the nuclear extracts were prepared and then subjected to non denaturizing polyacrylamide gel electrophoresis. The EMSA technique is based on the observation that protein-DNA complexes migrate slower than linear DNA in the gel matrix. Thus, the oligonucleotides with bound proteins are retarded in the gel and therefore are detectable as bandshifts, whereas unbound oligonucleotides run through the gel. To assess the specificity of the protein/DNA interaction, oligonucleotides with point mutations in the predicted binding motifs were used as controls.

Results

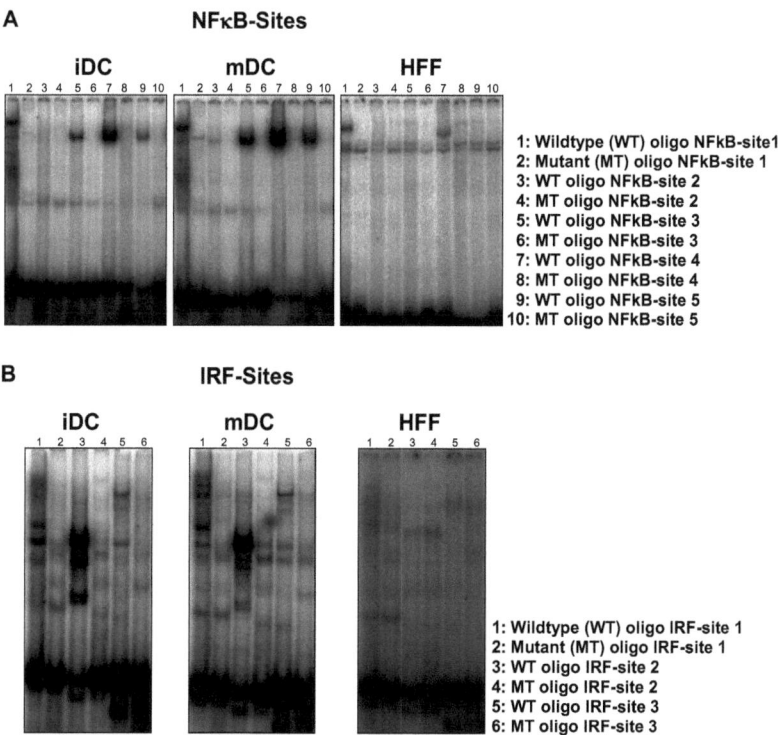

Figure 5.16. Analyses of the predicted NFκB-sites in iDCs, mDCs and HFF cells using EMSAs. (A) For EMSAs nuclear extracts derived from monocyte derived iDCs, 20h LPS-matured monocyte derived mDCs or HFF cells were incubated with ^{32}P-end-labeled double-stranded wildtype (lanes 1, 3, 5, 7, 9) or mutant (lanes 2, 4, 6, 8, 10) oligonucleotides coding for the predicted NFκB-sites and subsequently loaded onto a non reducing polyacrylamide gel. The radioactive signal was detected by a phospho image scanner at 635nm wavelength. One representative experiment out of three independently performed experiments is shown. (B) Analyses of the predicted IRF-sites in iDCs, mDCs and HFF cells using EMSAs. EMSAs were performed as described in (A) with wildtype (lanes 1, 3, 5, 7, 9) or mutant (lanes 2, 4, 6, 8, 10) oligonucleotides coding for the predicted IRF-sites. One representative experiment out of three independently performed experiments is shown.

Figure 5.16. (A) shows the EMSAs of the five predicted NFκB-sites within the UpP and the MP -261 sequence. The oligonucleotide coding for the NFκB-site 1 displayed a strong signal with nuclear extracts derived from iDCs, mDCs and HFF cells (lane 1), whereas in the mutant control no band was detectable (lane 2). Regarding the oligonucleotide coding for NFκB-site 2 only for mDCs a weak, but clear band was visible, whereas for iDCs and HFF cells the signal was absent (lanes 3 and 4). Thereby, this comparatively weak signal was in accordance with the biocomputational prediction of a weak TFBS within the UpP. For the NFκB-site 3 within the MP -261 a specific signal for iDCs and mDCs was detected, whereas HFF cells lacked this band (lanes 5 and 6). The NFκB-site 4 showed a specific signal for iDCs, mDCs and HFF cells (lanes 7 and 8). The oligonucleotide coding for the NFκB-site 5 revealed a strong signal for iDCs and mDCs, whereas no specific signal for HFF cells was observed (lanes 9 and 10).

Figure 5.16. (B) shows the EMSAs of the three predicted IRF-sites. None of the oligonucleotides coding for the IRF-sites showed a specific signal with nuclear extracts derived from HFF cells. The oligonucleotides coding for the IRF-sites 1 and 2 generated a specific signal for iDCs and mDCs (lanes 1 and 3), which was absent in the mutant controls (lanes 2 and 4). For the IRF-site 3 within the UpP also a specific, although weak, signal for iDCs and mDCs was detected, which was in accordance with the biocomputational prediction of a weak transcription factor binding site.

Taken together, all five predicted NFκB-sites, except of NFκB-site 2, bind nuclear factors present in both iDCs and mDCs. NFκB-site 2 only binds a nuclear factor from mDCs, but not from iDCs. Interestingly, only NFκB-sites 1 and 4 bind factors derived from HFF cells. However, all predicted IRF-sites bind factors present in both iDCs and mDCs, but not in HFF cells. Therefore, HFF cells were excluded from the following supershift experiments.

5.6.2. Supershift analyses for the transcription factors of the NFκB-family

While all five predicted NFκB-sites have been shown to induce characteristic shifts caused by specific protein binding to the target DNA, this change in mobility did not identify the bound protein in a shifted complex. Therefore, the identification of the protein bound to the probe is accomplished by including an antibody specific for the putative DNA-binding protein to the binding reaction. If the protein of interest binds to the target DNA, the antibody will bind to the protein-DNA complex, further decreasing its mobility relative to unbound DNA or protein-DNA complex without a bound antibody. This is also known as "supershift"-analyses. Thus, such assays were performed by the use of

Results

antibodies specific for single members of the human NFκB-family (Fig. 5.17. and 5.18., lanes 4-8). Figure 5.17. and 5.18. show the results where antibodies against p50 (lane 4), p65 (lane 5), cRel (lane 6), RelB (lane 7), p52 (lane 8) or the appropriate isotype control (lane 9) were applied. Several controls were included into these assays to ensure the specificity of the signal. Thus, one reaction only contained the radioactively labeled wildtype oligonucleotide without nuclear extract or antibody (lane 1) to warrant the purity of the probe (free oligo control). Lane 3 displays the reaction with a mutated oligonucleotide without the addition of an antibody to control the specificity of the signal. Two further specificity controls were included with (i) wildtype "*cold*" (unlabeled) competition oligonucleotides (lane 10) or (ii) mutant *cold* competition oligonucleotides (lane 11) instead of an antibody. The wildtype *cold* competition oligonucleotides compete with the radioactively labeled oligonucleotides for the protein binding and thus decrease the bandshift intensity. The mutated *cold* competition oligonucleotides do not compete with the radioactively labeled probe for the protein and therefore should not modulate the bandshift signal. Except of the addition of the antibodies or *cold* competition nucleotides, the experimental procedure was performed as described in 5.6.1.

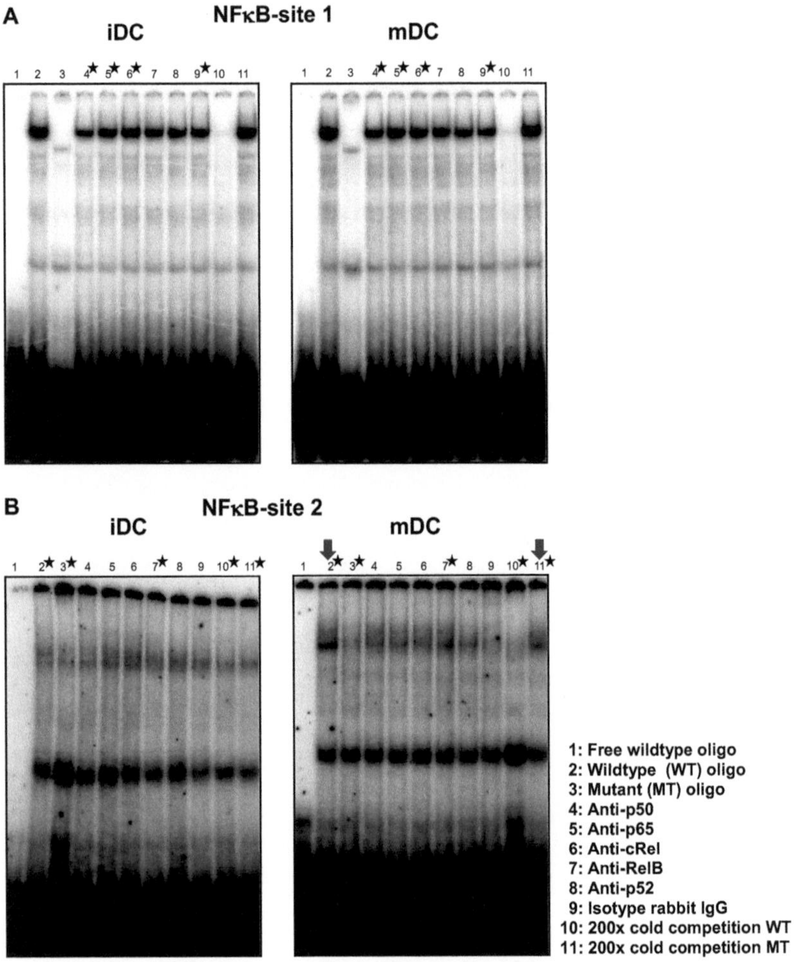

Results

A   NFκB-site 1

iDC   mDC

B   NFκB-site 2

iDC   mDC

1: Free wildtype oligo
2: Wildtype (WT) oligo
3: Mutant (MT) oligo
4: Anti-p50
5: Anti-p65
6: Anti-cRel
7: Anti-RelB
8: Anti-p52
9: Isotype rabbit IgG
10: 200x cold competition WT
11: 200x cold competition MT

Figure 5.17. The predicted NFκB-sites 1 and 2 show a bandshift in EMSAs, but no supershift. (A) and (B) For EMSAs nuclear extracts derived from either monocyte derived iDCs or 20 h LPS-matured monocyte derived mDCs were incubated with ^{32}P-end-labeled double-stranded wildtype (lanes 1, 3-11) or mutant (lane 2) oligonucleotides coding for the predicted NFκB-sites 1 and 2 and subsequently loaded onto a non reducing polyacrylamide gel. Lanes 3-11 show bandshift reactions that were incubated additionally with antibodies against the NFκB-family members p50, p65, cRel, RelB or p52 (lanes 4-8). Rabbit IgG isotype (lane 9), 200x molar excess of wildtype (lane 10) or mutant (lane 11) *cold* competition oligonucleotides have also been included as controls. The radioactive signal was detected by a phospho image scanner at 635nm wavelength. One representative experiment out of three independently performed experiments is shown. ⬇Supershifts, ★quantified and statistically evaluated signals.

Results

Figure 5.17. shows the supershift results obtained for the predicted NFκB-sites 1 and 2. For NFκB-site 1 (Fig. 5.17. A) a clear bandshift signal was observed in lane 2 with nuclear extracts derived from both iDCs and mDCs. The addition of an antibody, either against a member of the NFκB-family (lanes 4-8) or the isotype control (lane 9) did not alter the signal intensity nor cause a supershift for iDCs and mDCs, indicating that the bound protein was not a member of the NFκB-family.

For NFκB-site 2 (Fig. 5.17. B) no specific signal was observed in any of the reactions for iDCs, suggesting that no nuclear factor from iDCs binds to NFκB-site 2. However, a clear bandshift signal was observed in lane 2 for mDCs, strongly indicating the binding of a transcription factor from mDCs to this TFBs. Interestingly, lanes 4 to 9 showed a clearly decreased bandshift intensity in mDCs in comparison to the wildtype oligo in lane 2. This effect could not be attributed to a specific binding of an antibody to the bound transcription factor, as also the isotype control (lane 9) displayed a decreased bandshift intensity. One could speculated that the addition of an antibody, either directed against a member of the NFκB-family (lanes 4-8) or the isotype control (lane 9), unspecifically disrupted the binding of the protein to the oligonucleotide, suggesting a weak protein-DNA interaction.

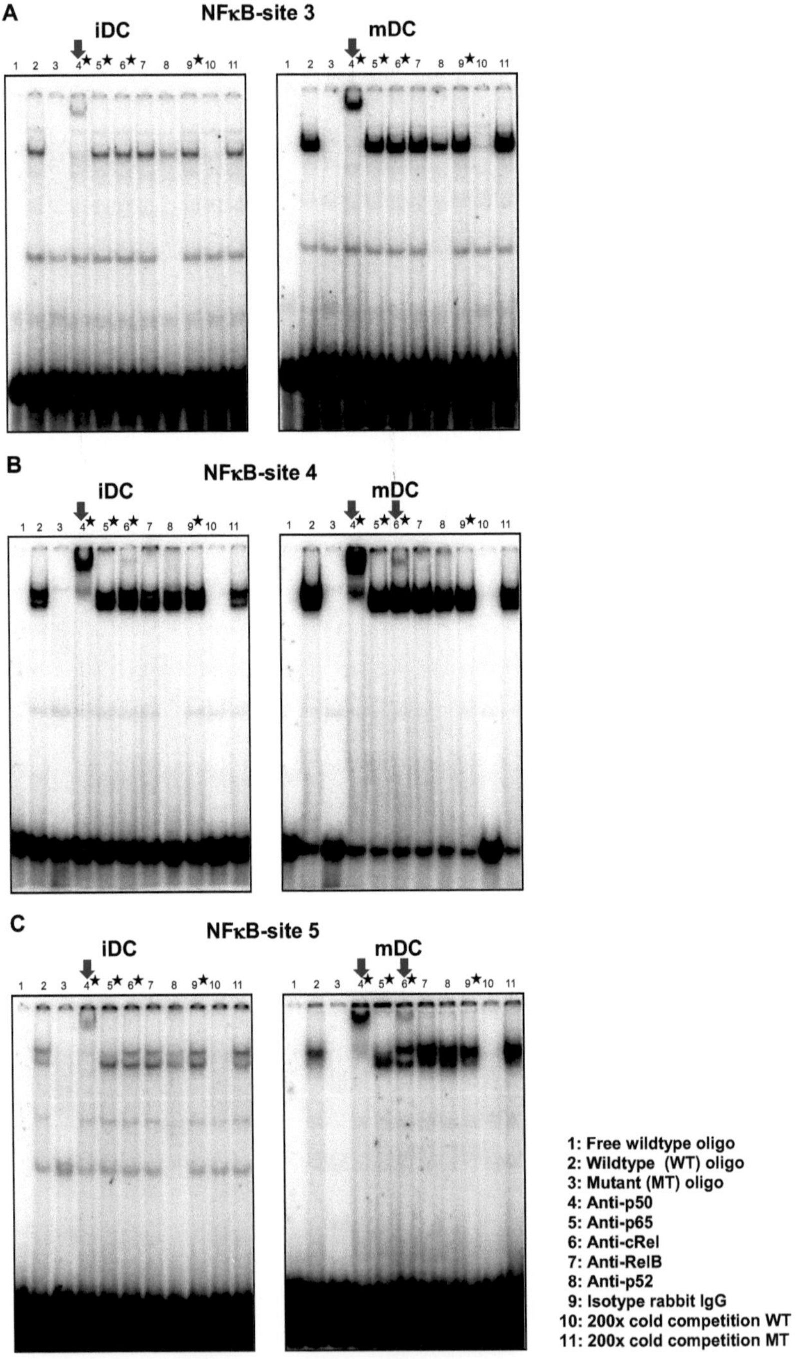
A
NFκB-site 3
iDC
mDC
1 2 3 4★ 5★ 6★ 7 8 9★ 10 11
1 2 3 4★ 5★ 6★ 7 8 9★ 10 11
B
NFκB-site 4
iDC
mDC
1 2 3 4★ 5★ 6★ 7 8 9★ 10 11
1 2 3 4★ 5★ 6★ 7 8 9★ 10 11
C
NFκB-site 5
iDC
mDC
1 2 3 4★ 5★ 6★ 7 8 9★ 10 11
1 2 3 4★ 5★ 6★ 7 8 9★ 10 11
1: Free wildtype oligo
2: Wildtype (WT) oligo
3: Mutant (MT) oligo
4: Anti-p50
5: Anti-p65
6: Anti-cRel
7: Anti-RelB
8: Anti-p52
9: Isotype rabbit IgG
10: 200x cold competition WT
11: 200x cold competition MT

Results

Figure 5.18. The predicted NFκB-sites 3, 4 and 5 show a bandshift and a supershift in EMSAs. (A)-(C) For EMSAs nuclear extracts derived from either monocyte derived iDCs or 20 h LPS-matured monocyte derived mDCs were incubated with ^{32}P-end-labeled double-stranded wildtype (lanes 1, 3-11) or mutant (lane 2) oligonucleotides coding for the predicted NFκB-sites 3, 4 and 5 and subsequently loaded onto a non reducing polyacrylamide gel. Lanes 3-11 show bandshift reactions that were incubated additionally with antibodies against the NFκB-family members p50, p65, cRel, RelB or p52 (lanes 4-8). Rabbit IgG isotype (lane 9), 200x molar excess of wildtype (lane 10) or mutant (lane 11) *cold* competition oligonucleotides have also been included as controls. The radioactive signal was detected by a phospho image scanner at 635nm wavelength. One representative experiment out of three independently performed experiments representing donors is shown. ↓ Supershifts, ★ quantified and statistically evaluated signals.

Figure 5.18. depicts the supershift results for the predicted NFκB-sites 3, 4 and 5. The experiments were performed as described previously for NFκB-sites 1 and 2. For all three analyzed NFκB-sites a clear supershift resulted from the addition of the anti-p50 antibody (lane 4) with nuclear extracts derived from both iDCs and mDCs. However, differences between iDCs and mDCs for each individual NFκB-site were observed: For NFκB-sites 3 and 5 (Fig. 5.18. A and C, respectively) the overall signal intensity appeared much weaker for iDCs than for mDCs. For NFκB-site 4 (Fig. 5.18. B) no difference in signal intensity between with iDCs and mDCs was observed. Additionally, for NFκB-sites 4 and 5 (Fig. 5.18 B and C, respectively), a slight supershift for cRel (lane 6) was observed for mDCs, but not for iDCs.

For all analyzed NFκB-sites the controls indicated a specific band- and supershift signal: The mutant controls in lane 3 did not show a bandshift signal. The wildtype *cold* competitions showed a loss of the signal, whereas the mutant *cold* competitions did not (lanes 10 and 11, respectively). Furthermore, no signal was observed in the presence of the free oligo control (lane 1), indicating the purity of the oligonucleotide probes.

Taken together, the nuclear factors binding to the predicted NFκB-site 1 and 2 could not be revealed by EMSAs. However, for the predicted NFκB-sites 3, 4 and 5 the anti-p50 antibody causes a supershift with nuclear extracts derived from iDCs and mDCs, indicating that p50 binds to the respective oligonucleotides. Additionally, the anti-cRel antibody causes a slight, but exclusive, supershift for the NFκB-sites 4 and 5 in mDCs, suggesting a weak binding of cRel to the respective sites.

Results

5.6.2.1. The statistical evaluation confirms the results of the EMSAs for the NFκB- sites

As shown above, EMSAs could not identify the nuclear factors binding to the predicted NFκB-sites 1 and 2. However, for NFκB-sites 3, 4, and 5, the NFκB-subunit p50 was revealed as binding factor, both in iDCs and mDCs. Additionally, for the NFκB-sites 4 and 5, NFκB-subunit cRel was observed as a binding transcription factor specifically in mDCs. In order to further validate the observations from EMSAs displayed in figure 5.17. and 5.18. in a statistical relevant manner, the underlying radioactive counts from three independently performed supershift experiments were quantified by the AIDA software and statistically evaluated with the Student's T-test and the Bonferroni-Holm *post hoc* test (Fig. 5.19.). For NFκB-sites 1, 3, 4 and 5, the intensities of the observed band- and supershifts of the reactions containing the anti-p50, anti-p65 and anti-cRel antibodies were quantified and then compared to the isotype control. The band- and supershift intensities of the controls were set to 1 and then the decrease in the bandshift intensity and the increase in the supershift intensity of the radioactively labeled oligonucleotides were quantified. For NFκB-site 2 (see Fig. 5.19. B), only the bandshifts with the wildtype and mutant control oligonucleotides without the addition of an antibody and the wildtype and mutant *cold* competition controls were compared and statistically evaluated, as the addition of an antibody unspecifically inhibited the formation of a bandshift (Fig. 5.17. B). Reactions that exhibited a significant statistical change in the band- or supershift in accordance with the previous visual observations from EMSA (see Fig. 5.17. and 5.18.) are marked with a filled arrow. Visually observed, but not statistical significant changes are marked with an empty arrow.

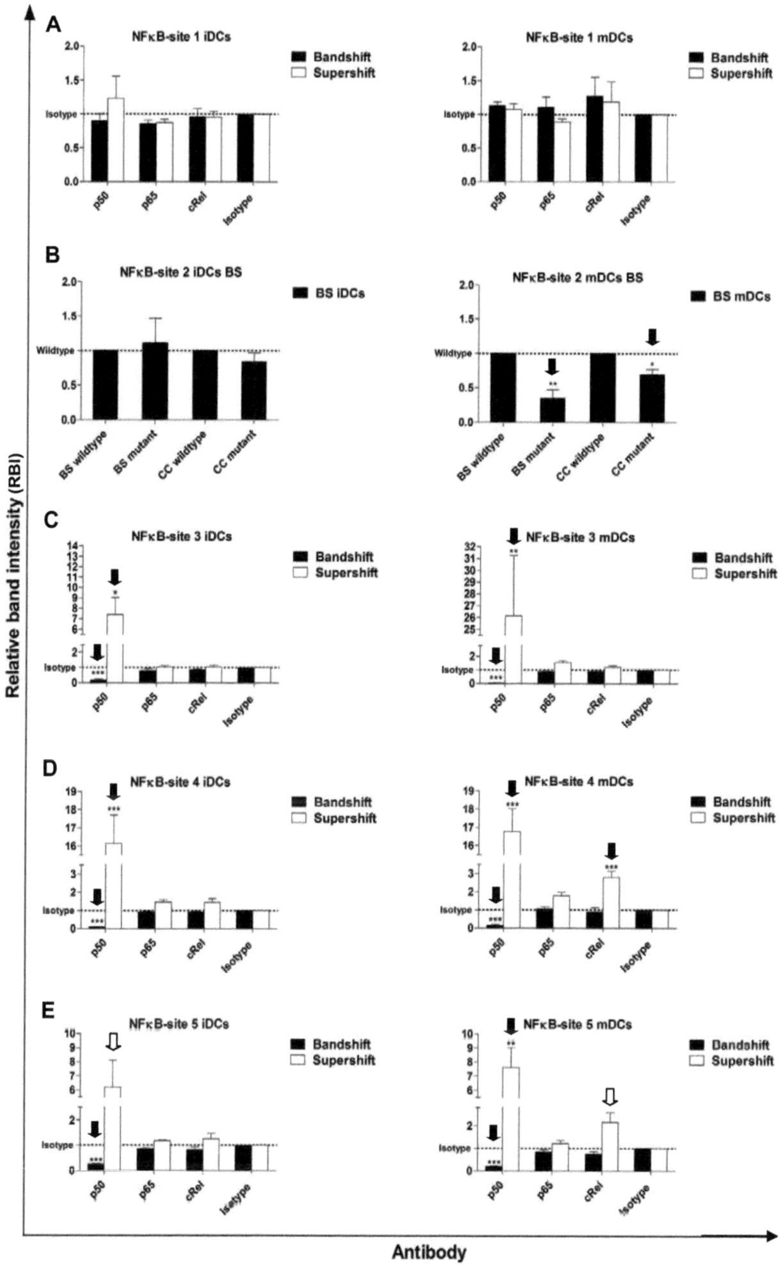
A
NFκB-site 1 iDCs
NFκB-site 1 mDCs
Bandshift
Supershift
Isotype
p50
p65
cRel
Isotype
B
NFκB-site 2 iDCs BS
BS iDCs
NFκB-site 2 mDCs BS
BS mDCs
Wildtype
BS wildtype
BS mutant
CC wildtype
CC mutant
C
NFκB-site 3 iDCs
NFκB-site 3 mDCs
D
NFκB-site 4 iDCs
NFκB-site 4 mDCs
E
NFκB-site 5 iDCs
NFκB-site 5 mDCs
Relative band intensity (RBI)
Antibody

Results

Figure 5.19. Statistical analyses of three NFκB EMSAs. (A)-(E) The bandshift and supershift signal intensities were quantified and compared to the isotype control using the AIDA software. The bandshift and the supershift of the isotype control were set to 1. An increase or decrease of the band intensity compared to the isotype control was statistically evaluated, except for the NFκB-site 2. For the NFκB-site 2 only the bandshifts with wildtype (WT) and mutant (MT) oligo without the addition of an antibody and the wildtype and mutant *cold* competitions (CCwt and CCmt, respectively) were compared and statistically evaluated. Filled red arrows indicate statistical significant band- or supershifts that are in concordance with the observations from chapter 5.6.2. The resulting p-values were determined via Student's T-test and Bonferroni-Holm *post hoc* test. * $p < 0.05$, ** $p < 0.025$, *** $p < 0.0166$. Results represent the means (±SEM) of three independently performed experiments. ⬇ Statistically confirmed band- or supershifts, ⬇visually observed, but not statistically significant supershifts.

For the NFκB-site 1 neither a decrease of the bandshift intensity nor a supershift were quantified as statistical significant in comparison to the isotype control (Fig. 5.19. A). NFκB-site 2 displayed significant changes in the bandshift intensities only for mDCs: the bandshift intensity of the mutant control oligonucleotide was significantly lower (0.35 RBI) than the bandshift intensity of the wildtype oligo (Fig. 5.19. B). Furthermore, the wildtype *cold* competition resulted in a significant decrease (0.7 RBI) of the bandshift intensity in comparison to the mutant control. Both controls indicated that the signal generated by the binding of a nuclear protein of mDCs is specific. For iDCs no significant differences between bandshift intensities of the wildtype oligonucleotides and the mutant controls were detected, thus indicating that no specific binding of a nuclear protein to the oligonucleotide occurred with nuclear extracts derived from iDCs. The NFκB-site 3 displayed a significant change in both band- and supershift intensity for the anti-p50 setup with nuclear extracts derived from both iDCs and mDCs in comparison to the isotype (Fig. 5.19. C). The overall effect was weaker in iDCs than when compared with mDCs. In iDCs the bandshift intensity decreased to 0.2 RBI and the supershift intensity increased to 7.4 RBI, whereas in mDCs the bandshift intensity decreased to 0.06 RBI and the supershift intensity increased to 26.1 RBI. In iDCs the bandshift intensity decreased to 0.1 RBI and while the supershift intensity increased to 16.1 RBI for the anti-p50 setting. In mDCs the bandshift intensity decreased to 0.1 RBI and the supershift intensity increased to 16.8 RBI. The NFκB-site 5 displayed a significant change in the bandshift for the anti-p50 setting with nuclear extracts derived from iDCs and mDCs (both 0.2 RBI; Fig. 5.19. E). The supershift was significant for mDCs (7.6 RBI), but not for iDCs (6.2 RBI). Additionally, a slight supershift in the anti-cRel setting was visible with nuclear extracts derived from mDCs, which was shown to be statistically significant for the NFκB-site 4 (2.8 RBI), but not for the predicted NFκB-site 5 (2.2 RBI).

Taken together, the quantification and statistical evaluation of the band- and supershift signals obtained from three independently performed experiments are in accordance with the visually observed results shown in figure 5.17. and 5.18.

5.6.3. Supershift analyses of the transcription factors of the IRF-family

To confirm the specific proteins binding to the predicted IRF-sites 1-3, again supershift assays were performed as described in chapter 5.6.2. As depicted in figure 5.20., antibodies specific for single members of the human IRF-family, IRF-1 to IRF-8 (lanes 4-11), or the appropriate isotype controls (lanes 12 and 13) were added to the reaction mix. Further controls included: (i) Free oligo control (lane 1), (ii) mutant control (lane 3) and (iii) *cold* competition control, either wildtype or mutant (lanes 14 and 15, respectively).

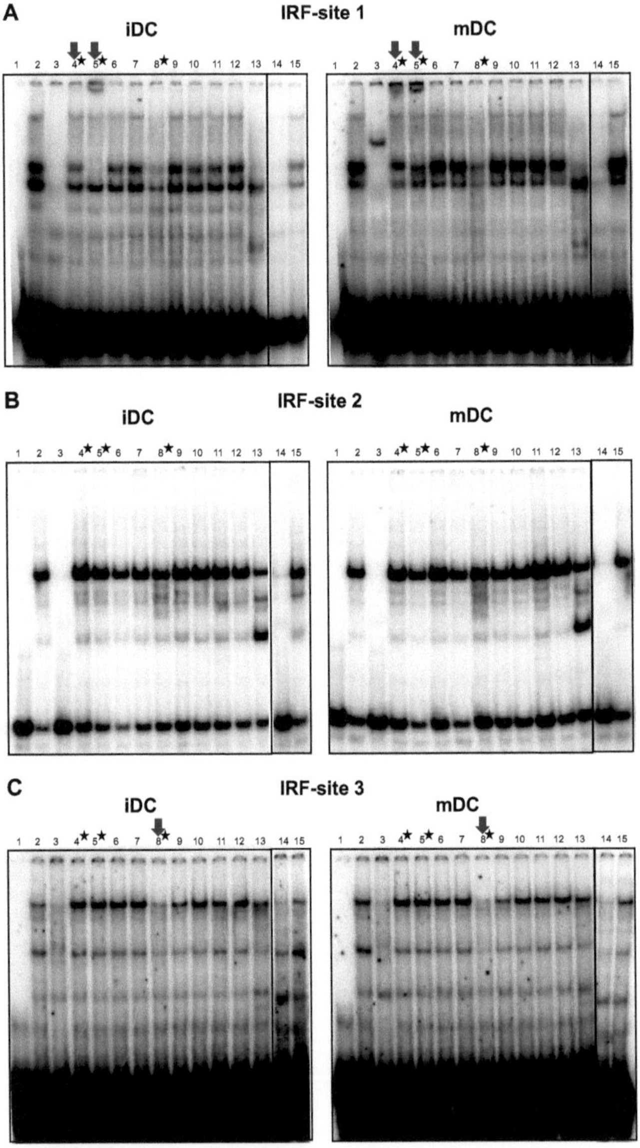
A
IRF-site 1
iDC
mDC
B
IRF-site 2
iDC
mDC
C
IRF-site 3
iDC
mDC
1 2 3 4★ 5★ 6 7 8★ 9 10 11 12 13 14 15

1: Free wildtype oligo
2: Wildtype (WT) oligo
3: Mutant (MT) oligo
4: Anti-IRF-1
5: Anti-IRF-2
6: Anti-IRF-3
7: Anti-IRF-4
8: Anti-IRF-5
9: Anti-IRF-6
10: Anti -RF-7
11: Anti-IRF-8
12: Isotype rabbit IgG
13: Isotype goat IgG
14: 200x cold competition WT oligo
15: 200x cold competition MT oligo

Results

Figure 5.20. All predicted IRF-sites show a bandshift in EMSAs, but only sites 1 and 3 were confirmed by supershift. (A)-(C) For EMSAs nuclear extracts derived from either monocyte derived iDCs or 20 h LPS-matured monocyte derived mDCs were incubated ^{32}P-end-labeled double-stranded wildtype (lanes 1, 3-15) or mutant (lane 2) oligonucleotides coding for the predicted IRF-sites 1, 2 and 3. Lanes 3-11 show bandshift reactions that were incubated additionally with antibodies against the members of the IRF-family (IRF-1 to IRF-8; lanes 4-11). Rabbit or goat IgG isotypes (lane 12 and 13, respectively), 200x molar excess of wildtype (lane 14) or mutant (lane 15) cold competition oligonucleotides have also been included as controls. The radioactive signal was detected by a phospho image scanner at 635nm wavelength. One representative experiment out of three independently performed experiments is shown. ↓Supershifts, ★quantified and statistically evaluated signals.

IRF-site 1 showed a strong bandshift signal for both iDCs and mDCs (lane 2; Fig. 5.20. A). The addition of an anti-IRF-1 antibody to the nuclear extracts derived from iDCs caused a slight decrease of the bandshift signal intensity (lane 4) and a weak supershift in comparison to the rabbit isotype control (lane 12). In contrast, the addition of an anti-IRF-2 antibody caused a strong decrease of the bandshift signal intensity (lane 5), as well as a substantial supershift (lane 12) for iDCs. Conversely, to the effects observed for iDCs, the anti-IRF-1 antibody induced a clear supershift with nuclear extracts derived from mDCs together with a slight decrease of the bandshift signal intensity (lane 4). The addition of an anti-IRF-2 antibody again caused a clear supershift for mDCs, but only a slight decrease of the bandshift intensity in comparison to iDCs (lane 5). The anti-IRF-5 antibody caused a slight decrease of the bandshift signal intensity (lane 8) for iDCs and mDCs, but no supershift. The same was observed for the goat isotype control (lane 13), indicating that this effect was not specific. With all other antibodies (IRF-3, 4 and IRF-6 to IRF-8) neither for iDCs nor for mDCs supershifts or differences in the bandshift intensities were observed.

Although a strong bandshift signal was observed in lane 2 for iDCs and mDCs, no supershift or decrease of bandshift intensity was observed for IRF-site 2 for all used antibodies of the IRF-family members (lanes 4-11), for both iDCs and mDCs (Fig. 5.20. B). This indicated that the bound protein was not a member of the IRF-family. IRF-site 3 showed a clear bandshift signal (lane 2) for both iDCs and mDCs that decreased with the addition of the IRF-5 antibody for both cell types (lane 8) in comparison to the goat isotype control (lane 13; Fig. 5.20. C). The decrease of the bandshift intensity with the anti-IRF-5 antibody was stronger for mDCs than for iDCs. Furthermore, no supershift was observed with the anti-IRF-5 antibody for both cell types. No further supershifts or decrease in bandshift intensities were observed for all other anti-IRF antibodies for iDCs or mDCs.

For all three analyzed IRF-sites the mutant control in lane 3 did not exhibit a bandshift signal. The wildtype *cold* competition showed a loss of a signal, whereas the mutant *cold* competition did not (lanes 14 and 15, respectively). Furthermore no signal was generated with the free oligo control (lane 1). Thus, the controls indicated the purity of the probes as well as the specificity of the bandshift signals.

Taken together, for the predicted IRF-site 1, the anti-IRF-1 and the anti-IRF-2 antibodies caused a decrease of the bandshift intensity and an increase of the supershift intensity for iDCs and mDCs, whereby the effects for IRF-1 are more prominent for mDCs. In contrast, the effects for IRF-2 are stronger for iDCs than for mDCs. This indicates that IRF-1 and IRF-2 differentially bind to the respective oligonucleotides in EMSAs, depending on the maturation status of the DCs. Whereas no specific factor for IRF-site 2 could be identified, the anti-IRF-5 antibody induced a decrease of the bandshift signal for IRF-site 3, but no supershift, for both iDCs and mDCs. This indicates that IRF-5 binds to the respective oligonucleotide in EMSAs, but is specifically inhibited by the antibody.

5.6.3.1. The statistical evaluation confirms the results of the EMSAs for the IRF-sites

EMSAs indicated that IRF-1 and IRF-2 bind in a maturation status dependent manner to IRF-site 1 in DCs. Moreover, IRF-5 was revealed as a strongly binding factor at IRF-site 3 for mDCs and to a weaker extent for iDCs. However, EMSAs could not identify the nuclear factors binding to the predicted IRF-sites 2.

As for the NFκB-sites (see chapter 5.6.1.2.), statistical validation of the results from EMSAs were performed by quantification of the underlying radioactive counts from four independently performed experiments by the AIDA software and statistical evaluation with Student's T-test and Bonferroni-Holm *post hoc* test (Fig. 5.21.). For IRF-sites 1-3, the intensities of the observed band- and supershifts of the reactions containing the anti-IRF-1, anti-IRF-2 and anti-IRF-5

Results

antibodies were quantified and then compared to the respective isotype control. The band- and supershift intensities of the isotype controls were set to 1 and the decrease in the bandshift intensity and the increase in the supershift intensity of the radioactively labeled oligonucleotides were quantified. Reactions that exhibited a statistical change in the band- or supershift in accordance with the previous EMSA results (see Fig. 5.20.) are marked with a filled arrow. Visually observed, but not statistical significant changes are marked with an empty arrow.

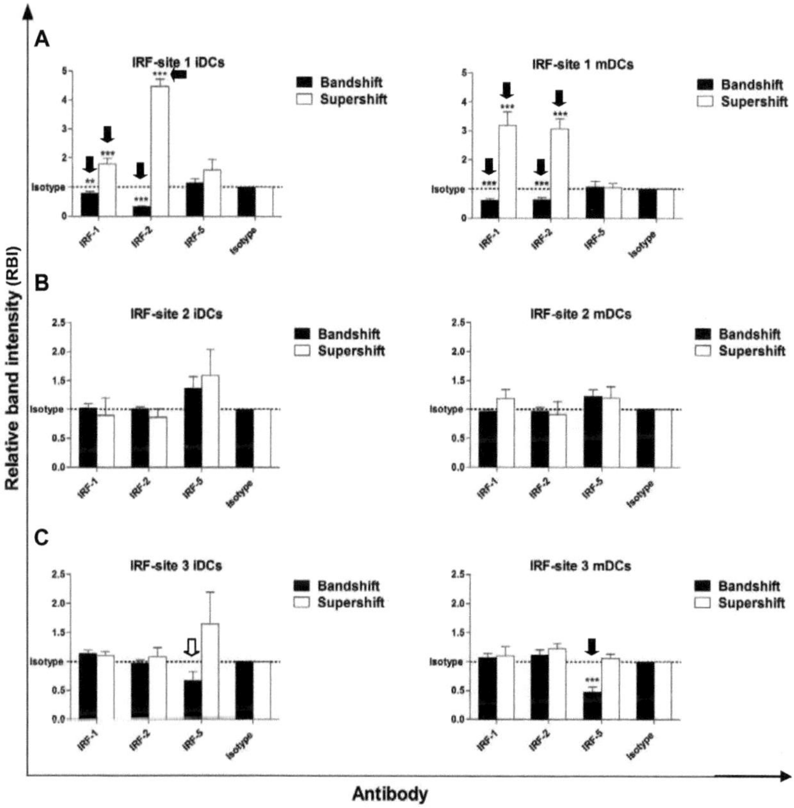

Figure 5.21. Statistical analyses of three IRF EMSAs (A)-(C). The bandshift and supershift signal intensities were quantified and compared to the isotype control using the AIDA software. The bandshift and the supershift of the isotype control were set to 1. An increase or decrease of the band intensity compared to the isotype control was statistically evaluated. The resulting p-values were determined via Student's T-test and the Bonferroni-Holm *post hoc* test. * $p < 0.05$, ** $p < 0.025$, *** $p < 0.0166$. Results represent the means (±SEM) of three independently performed experiments from different donors. ↓ Statistically confirmed supershifts, ⇩ visually observed, but not statistically significant supershifts.

The predicted IRF-site 1 showed a significant decrease of the bandshift intensity and a significant increase of the supershift intensity for the anti-IRF-1 and the anti-IRF-2 setting for both iDCs and mDCs (Fig. 5.21 A). Concerning anti-IRF-1, the effect for iDCs was weaker than for mDCs. For iDCs the bandshift intensity decreased to 0.8 RBI and the supershift intensity increased to 3.2 RBI, whereas for mDCs it decreased to 0.6 RBI and increased to 3.2 RBI, respectively. In contrast, the effect for the anti-IRF-2 setting was somewhat stronger for iDCs than for mDCs. For iDCs the bandshift intensity decreased to 0.3 RBI (0.6 RBI for mDCs) and the supershift intensity increased to 4.5 RBI (3.0 RBI for mDCs). For the predicted IRF-site 2 neither a decrease of the bandshift intensity nor a supershift with any of the examined antibodies were assessed as statistical significant in comparison to the isotype controls (Fig. 5.21 B). For the predicted IRF-site 3, only the bandshift intensity for mDCs decreased significantly to 0.5 RBI in comparison to the isotype control. A supershift was neither observed nor statistically significant (Fig. 5.21 C).

In summary, the quantification and statistical evaluation of the band- and supershift signals obtained from three independently performed experiments are in accordance with the results in figure 5.20.

5.6.4. Summary of the EMSAs

Taken together, it could be shown that the biocomputationally predicted IRF- and NFκB-sites are bound by nuclear factors. Supershift experiments verify three of five predicted NFκB-sites by binding a member of the respective protein family. Moreover, two of three predicted IRF-sites are not only shown to bind nuclear factors, but also to bind members of the IRF-family in supershift experiments. Interestingly, in some cases the binding of the nuclear factors is cell type- and maturation status-dependent. Figure 5.22. summarizes the results from the EMSAs and the statistical analyses: (i) IRF-5 binds weakly to the IRF-site 3 at the UpP in iDCs. The factor that binds to the NFκB-site 1 is not identified yet and no binding of a nuclear factor could be observed for NFκB-site 2. At the MP -261, p50 binds weakly to the NFκB-sites 3 and 5 and strongly to the NFκB-site 4. At the *185 bp enhancer* the IRF-site 1 strongly binds IRF-2 and weakly IRF-1. The factor binding to the IRF-site 2 could not be disclosed by

EMSAs. (ii) In contrast, in mDCs IRF-5 binds strongly to the IRF-site 3 in the UpP. Furthermore, not yet revealed factors bind to the NFκB-sites 1 and 2. At the MP -261, p50 strongly binds to the NFκB-sites 3, 4 and 5. Moreover, cRel binds weakly to the NFκB-sites 4 and 5. In contrast to iDCs, IRF-1 binds strongly to the IRF-site 1 in the *185 bp enhancer*. The nuclear factor binding to the IRF-site 2 is not revealed yet. (iii) As for HFF cells, the binding of an unidentified nuclear factor has been observed for NFκB-sites 1 and 3, but not for the other transcription factor binding sites.

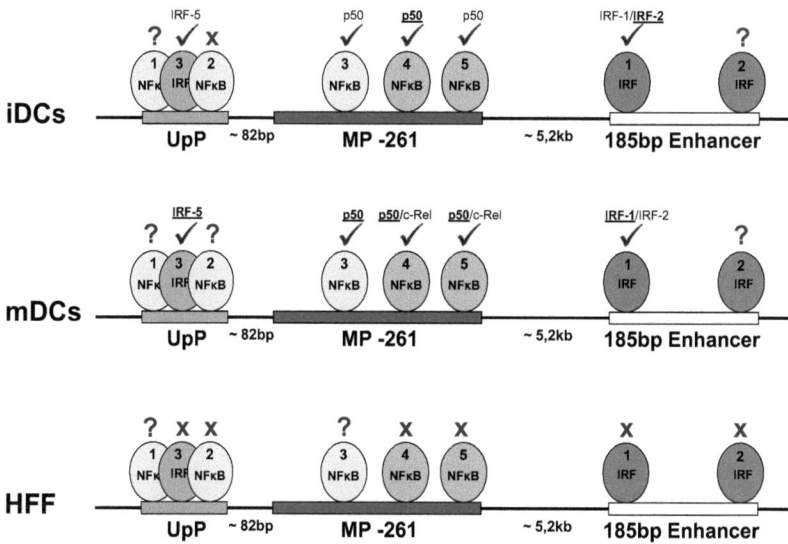

Figure 5.22. Schematic depiction of the results obtained from EMSAs. Factors bound to NFκB- and IRF-sites are depicted as ovals. Transcription factors in underlined bold type indicate a stronger signal intensity than the corresponding factor in a different cell type. ✓ Verified binding site, **X** no binding, **?** binding by a not yet identified factor.

Results

5.7. Functional analyses of the three IRF-sites using mutagenesis and luciferase assays

According to the bioinformatical model (Fig. 5.10.) the three predicted IRF-sites are crucial elements for the regulation of CD83 expression. EMSAs revealed that members of the IRF-family bind to the IRF-sites 1 and 3 and a yet unknown nuclear factor to the predicted IRF-site 2. However, no functional evidence was provided so far. In order to investigate the involvement of the predicted UpP and the impact of the three IRF-sites individually, luciferase reporter constructs with mutated IRF-sites were performed (Fig. 5.23.).

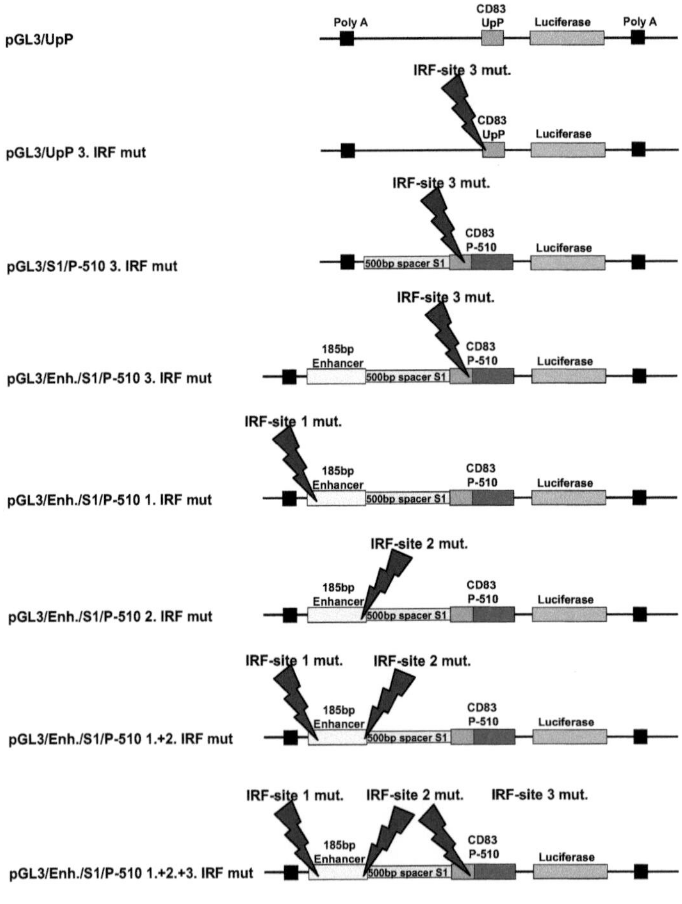

160

Results

5.23. Schematic depiction of reporter constructs containing mutated IRF-sites used for luciferase assays. To generate the pGL3/UpP plasmid, a 244 bp fragment of the P -510 containing the UpP was isolated by enzymatic digestion and cloned upstream of the luciferase gene into the pGL3/Basic plasmid. The individual IRF-sites in the pGL3/UpP, the pGL3/S1/P-510 and the pGL3/185 bp Enh./S1/P-510 plasmids were mutated by PCR mutagenesis. ◄ Mutated IRF-site.

Based on the previously generated plasmids (see Fig. 5.11.) twelve plasmids with different combinations of mutated IRF-sites were generated. Therefore, IRF-sites 1-3 in pGL3/S1/P-510 and pGL3/185bp Enh./S1/P-510 were mutated by the introduction of four (IRF-sites 1 and 3) or five (IRF-site 2) point mutations by PCR mutagenesis. The point mutations of the IRF-sites matched the mutant control oligonucleotides from EMSAs (see chapter 5.6.). Furthermore a plasmid was generated where the luciferase gene is driven by the isolated wild type UpP (pGL3/UpP) as well as the mutated UpP (pGL3/UpP 3. IRFmut). Therefore, a 244 bp fragment containing the UpP was enzymatically digested from the pGL3/P -510 plasmid, either mutated in the IRF-site 3 or not and cloned upstream of the luciferase gene into the pGL3/Basic plasmid.

5.7.1. Mutation of individual IRF-sites results in a significantly reduced luciferase expression in XS52 cells and mDCs

In order to determine the functionality of the three IRF-sites in the UpP and the *185 bp enhancer*, XS52 cells were transfected with pGL3/S1/P-510 containing the enhancer or not, the pGL3/UpP or their mutated counterparts (Fig. 5.23.; Fig. 5.24. A and B). The induction of the P -510 (Fig. 5.24. A) or the UpP (Fig. 5.24. B) was compared to the corresponding plasmids with different combinations of mutated IRF-sites by a luciferase assay 48 h after transfection.

Results

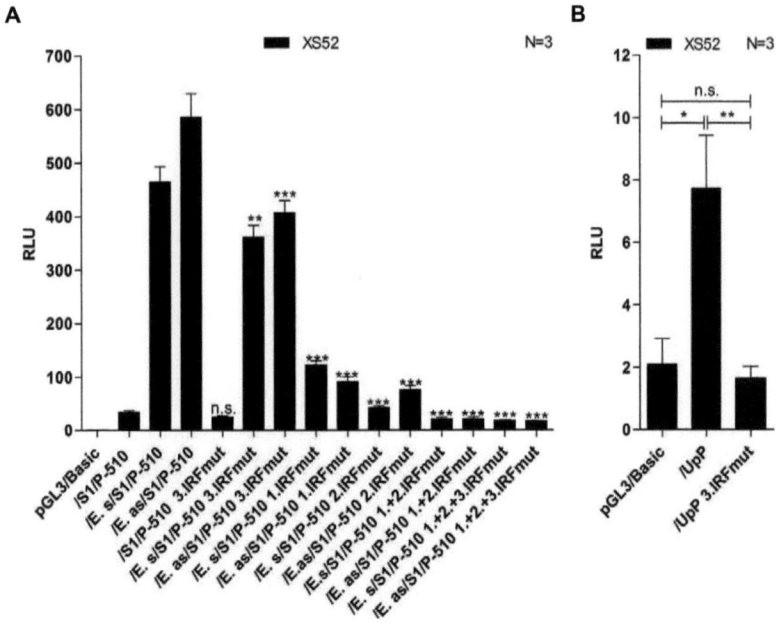

Figure 5.24. Mutation of any of the three IRF-sites in the ternary complex significantly reduces the luciferase expression in XS52 cells. (A) and (B) XS52 cells were transfected with 2.5 μg of the wildtype pGL3/UpP or the pGL3/S1/P -510 with and without the *185 bp enhancer* in sense (s) or antisense (as) orientation upstream of the P -510 or with the analogical plasmids containing one, two or three mutated IRF-sites. The promoter activity was determined by luciferase reporter assays 48 h after transfection. As controls the pGL3/Basic and the pGL3/CMV/luc vector were transfected to assess background activity and for internal normalization, respectively. The resulting RLU were normalized to the protein concentration of each lysate. The plasmids containing the different mutations were compared to the respective wildtype control and the resulting p-values were determined via one way ANOVA and Bonferroni's Multiple Comparison *post hoc* test. The mutated IRF-sites are indicated by their respective numbers. * $p < 0.05$, ** $p < 0.01$, *** $p < 0.001$ and n.s. not significant (p>0.05). Results represent the means (±SEM) of three independently performed experiments.

As depicted in figure 5.24. (A), the background activity of the pGL3/Basic vector in XS52 amounted to 1.2 RLU. The basal activity of the P -510 (26.2 RLU) was strongly induced by the *185 bp enhancer* in sense (436.5 RLU) and antisense (560.2 RLU) orientation. The mutation of IRF-site 3 (pGL3/S1/P-510/ 3.IRFmut) did not affect the basal activity of the P -510 significantly (25.1 RLU). However, in combination with the *185 bp enhancer*, either in sense or antisense direction, the induction of the P -510 containing the mutated IRF-site 3 was significantly reduced to 356.9 RLU (pGL3/E.s/S1/P-510 3.IRFmut) and 414.3 RLU

Results

(pGL3/E.as/S1/P-510 3.IRFmut). The single mutation of the IRF-site 1 in the *185 bp enhancer*, regardless of their orientation, resulted in a drastic loss of P -510 activity with 111.4 in sense and 81.4 RLU in antisense orientation. For the single mutation of IRF-site 2 in the *185 bp enhancer* the remaining induction of the P -510 was comparable to that resulting from the mutation of IRF-site 1 (sense: 40.0 RLU; antisense: 70.0 RLU). The combined mutation of IRF-sites 1 and 2 in the enhancer or of all three IRF-sites completely abolished the induction of P -510, regardless of the orientation of the enhancer. With both mutated IRF-sites in the enhancer (IRF-site 1 and 2), the remaining RLU amounted to 21.4 RLU in sense and 20.7 RLU in antisense orientation, with all three IRF-sites mutated, to 17.9 RLU in sense and 16.5 RLU in antisense orientation. Figure 5.24. (B) shows that the basal activity of the UpP was significantly over background (7.7 RLU) in comparison to pGL3/Basic (2.1 RLU). Upon mutation of IRF-site 3 (pGL3/UpP 3.IRFmut), the basal activity dropped significantly down to background level (1.6 RLU), outlining the important role of the IRF-site 3. In comparison to the full length P -510, the basal activity of the UpP was 3.4 fold under the basal activity of the P -510.

To confirm the functionality of all three IRF-sites also in monocyte derived DCs, iDCs were electroporated with the same plasmids depicted in figure 5.23. (except pGL3/UpP and pGL3/UpP 3.IRFmut). Three hours after electroporation cells were matured for 20 h with LPS and the induction of the P -510 was assessed by a luciferase reporter assay (Fig. 5.25.).

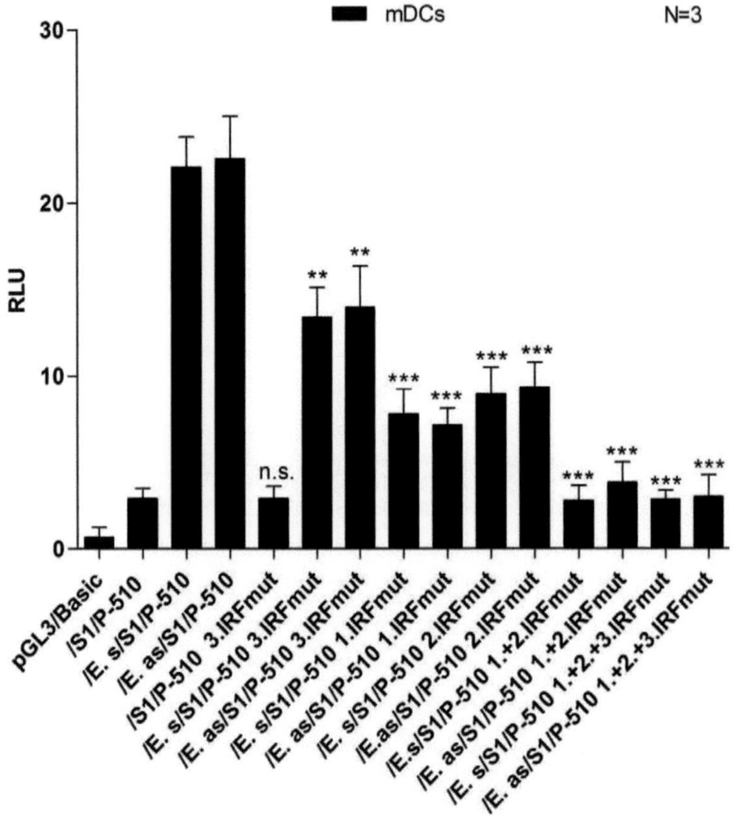

Figure 5.25. Mutation of any of the three IRF-sites in the ternary complex significantly reduces the luciferase expression in mDCs. Monocyte derived iDCs were electroporated with 4 µg of either wildtype pGL3/S1/P -510 with and without the *185 bp enhancer* in sense (s) or antisense (as) orientation upstream of the P -510 or with the analogical plasmids containing one, two or three mutated IRF-sites. Cells were matured 3 h after electroporation with LPS to a final concentration of 0.1 ng/ml. The promoter activity was determined by luciferase reporter assays 20 h after LPS addition. As controls the pGL3/Basic and the pGL3/CMV/luc vector were transfected to assess background activity and for internal normalization, respectively. The resulting RLU were further normalized to the protein concentration of each lysate. The plasmids containing the different mutations were compared to the respective wildtype control and the resulting p-values were determined via one way ANOVA and Bonferroni's Multiple Comparison *post hoc* test. The mutated IRF-sites are indicated by their respective numbers. * $p < 0.05$, ** $p < 0.01$, *** $p < 0.001$ and n.s. not significant ($p > 0.05$). Results represent the means (±SEM) of three independently performed experiments from different donors.

In mDCs the background activity of the pGL3/Basic vector amounted to 0.8 RLU. The basal activity of the non mutated P -510 (2.9 RLU) was strongly induced by the *185 bp enhancer* in both orientations (22.0 and 22.5 RLU in sense and antisense orientation, respectively; Fig. 5.24. C). Without the *185 bp enhancer* the mutation of IRF-site 3 in the P -510 did not significantly affect the basal activity of P -510 (2.9 RLU), whereas in the presence of the enhancer, the induction was significantly reduced. With the *185 bp enhancer* in sense orientation the remaining induction resulted in 13.4 RLU and in antisense orientation in 14.0 RLU. Mutations of both IRF-sites in the *185 bp enhancer* significantly reduced induction of P -510, regardless of the orientation of the enhancer. The mutation of IRF-site 1 resulted in 7.8 RLU (sense) and 7.2 RLU (antisense). For the mutation of IRF-site 2 in the *185 bp enhancer* the remaining induction of P -510 was comparable to that resulting from the mutation of IRF-site 1, amounting to 9.0 RLU (sense) and 9.4 RLU (antisense). The combined mutation of IRF-sites 1 and 2 in the enhancer or of all three IRF-sites completely abolished the induction of the P -510, regardless of the orientation of the enhancer. With both mutated IRF-sites in the enhancer (IRF-site 1 and 2), the RLU amounted to 2.7 RLU in sense and 3.8 RLU in antisense orientation. After mutation of all three IRF-sites, RLU added up to 2.8 in sense and 3.0 in antisense orientation.

Taken together, all three IRF-sites significantly influence the CD83 promoter activity in XS52 cells and mDCs, since any individual mutation resulted in a drastically reduced induction of luciferase expression. Hence, these results provide strong functional evidence for the validity of the bioinformatical model regarding the formation of a ternary complex consisting of UpP, MP -261 and *185 bp enhancer*.

5.8. Verification of the functionality of the NFκB-sites in UpP and MP -261 by cotransfection experiments in 293T cells

EMSAs revealed the binding of IRF-5, p50 and cRel to the oligonucleotides coding for the IRF-site 3 in the UpP and for the NFκB-sites in the MP -261, respectively. Although p65 did not bind in EMSAs to any of the oligonucleotides, the transcription factor was nevertheless taken into consideration for the following cotransfection experiments, since p50 and p65 form the classical NFκB heterodimer and Berchtold and colleagues reported the binding of p65 to the MP -261 in 2002[167]. In order to analyze the effects of IRF-5, p50, p65 and cRel on the induction of the MP -261 (Fig. 5.27.) and the UpP (Fig. 5.28.), luciferase reporter plasmids containing one of these promoters were cotransfected with different combinations of plasmids coding for IRF-5, p50, p65 and cRel.

5.8.1. IRF-5, p50, p65 and cRel can be overexpressed in 293T cells

To start with, the successful overexpression of IRF-5, p50, p65 and cRel after transfection of 293T cells was assessed by Western blot analyses (Fig. 5.26.). Therefore, 293T cells were transfected either with the pCDNA3.1 expression vector coding for IRF-5, p50, p65 and cRel or the empty vector backbone using. The next day, cells were harvested and lysed in RIPA buffer for the generation of whole cell lysates. Protein expression was assessed using Western blot analyses by staining with anti-human IRF-5, p50, p65 or cRel primary antibodies.

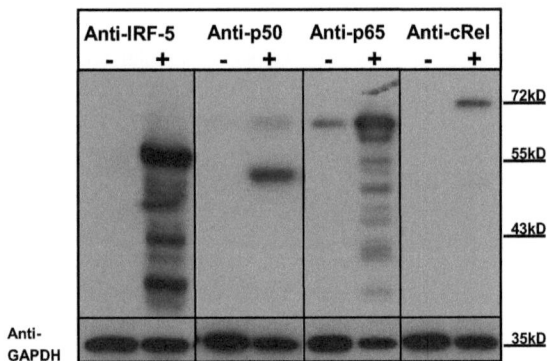

Figure 5.26. IRF-5, p50, p65 and cRel are efficiently overexpressed in 293T cells. 293T cells were transfected with 0.5 µg of total DNA consisting of either 0.15 µg of the pCDNA3.1 plasmid coding for IRF-5, p50, p65 or cRel and 0.35 µg pCDNA3.1 vector backbone (labeled +) or 0.5 µg pCDNA3.1 vector backbone alone as control (labeled -). For the generation of whole cell lysates cells were lysed 48 h after transfection. For Western blot analyses the nitrocellulose membranes were stained with a goat anti-IRF-5 or rabbit anti-p50, p65 or cRel and a mouse anti-human GAPDH antibody (loading control). The signal was detected via chemiluminescence that was visualized on a photo film. One representative experiment out of three independently performed experiments is shown.

Results

As shown in figure 5.26., after transfection with the vector backbone no signal for IRF-5, p50, p65 or cRel could be detected in 293T cells. However, transfection of pcDNA3.1 vector coding for IRF-5, p50, p65 and cRel generated a strong signal at the expected molecular weight (~58kDa, ~50kDa, ~65 kDa and ~72kDa, respectively).

Taken together, IRF-5, p50, p65 and cRel can be successfully overexpressed in 293T cells.

5.8.2. IRF-5, p50, p65 and cRel induce the MP -261 in 293T cells

EMSAs revealed a binding of NFκB subunits p50 and cRel to the respective TFBs within the MP -261. Moreover, Berchtold and colleagues reported a binding of NFκB-subunit p65 to NFκB-site 4 in the MP -261[167]. However, also IRF-5 might play a role in the induction of the MP -261, as Krausgruber and colleagues reported recently the cooperative binding of IRF-5 and cRel to a NFκB binding element in the TNF-α promoter[425]. Furthermore, IRF-5 was shown to bind to IRF-site 3 in the UpP using EMSAs indicating an involvement in the transcriptional regulation of CD83. To validate the data obtained by EMSAs in a functional assay, 293T cells were cotransfected with the pGL3/MP -261 reporter plasmid and different combinations of the pCDNA3.1 expression vector coding for p50, p65, cRel or IRF-5, or empty pCDNA3.1 vector as control (Fig. 5.27.). Forty eight hours after transfection the induction of the MP -261 was assessed by luciferase assays.

Results

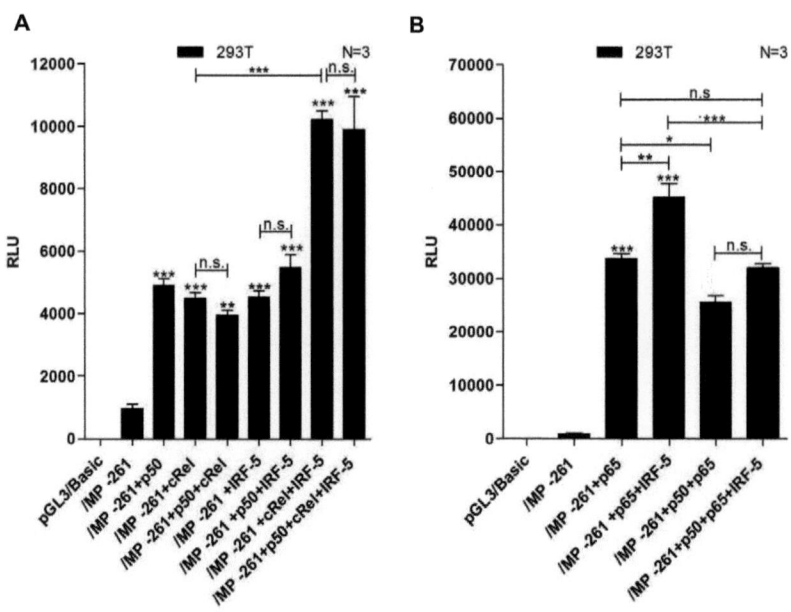

Figure 5.27. Transcription factors of the NFκB-family and IRF-5 induce the MP -261 in 293T cells. (A) and (B) 293T cells were cotransfected with 0.05 µg of pGL3/MP -261 reporter plasmid and 0.15 µg of each plasmid coding for the indicated transcription factors (p50, p65, cRel or IRF-5 in pCDNA3.1) or empty vector backbone as control. The total amount of DNA was adjusted with pCDNA3.1 empty vector backbone to 0.5 µg. The promoter activity was determined by luciferase reporter assays 48 h after transfection. As controls the pGL3/Basic and the pGL3/CMV/luc vector were transfected to assess background activity and for internal normalization, respectively. The resulting RLU were normalized to the protein concentration of each lysate. The cotransfections were compared to the pGL3/MP -261 control and the resulting p-values were determined via one way ANOVA and Bonferroni's Multiple Comparison *post hoc* test. * $p < 0.05$, ** $p < 0.01$, *** $p < 0.001$ and n.s. not significant (p>0.05). Results represent the means (±SEM) of three independently performed experiments.

The pGL3/Basic plasmid displayed a background activity of 0.8 RLU in 293T cells (Fig. 5.27. A). The basal activity of the MP -261 alone amounted to 966.2 RLU. However, cotransfection with pCDNA3.1 plasmids coding for p50, cRel or IRF-5 or combinations of these plasmids led to a significant induction of the MP -261. Cotransfection with pCDNA3.1/p50 induced the MP -261 to 4919.5 RLU. The cotransfection with pCDNA3.1 coding either for cRel or IRF-5 induced the MP -261 to a comparable extend (4490.0 RLU and 4532.9 RLU, respectively). Interestingly, neither the dual cotransfection with pCDNA3.1/p50 and pCDNA3.1/cRel (3942.8 RLU) nor with pCDNA3.1/p50 and pCDNA3.1/IRF-5

Results

(5479.4 RLU) induced the MP -261 stronger than with the single plasmids. However, the cotransfection with pCDNA3.1/cRel in combination with pCDNA3.1/IRF-5 induced the MP -261 in an additive manner (10222.5 RLU), thus significantly stronger than pCDNA3.1/cRel or pCDNA3.1/IRF-5 alone. The triple cotransfection with pCDNA3.1/cRel, pCDNA3.1/IRF-5 and pCDNA3.1/p50 did not significantly alter the induction of the MP -261 (9904.8 RLU) in comparison to the dual combination of pCDNA3.1/cRel and pCDNA3.1/IRF-5. The strongest induction of the MP -261 in 293T cells resulted from cotransfections containing p65 (5.27. B). Cotransfection with pCDNA3.1/p65 induced the MP -261 stronger than any other transcription factor alone (33738.7 RLU). The induction could be significantly raised to 45192.1 RLU by dual cotransfection with pCDNA3.1/p65 and pCDNA3.1/IRF-5. Interestingly, the dual cotransfection using pCDNA3.1/p65 and pCDNA3.1/p50 significantly reduced the induction of the MP -261 (25611.8 RLU) in comparison to the cotransfection with pCDNA3.1/p65 alone. A similar significant reduction in the induction of the MP -261 was observed with the triple combination with pCDNA3.1/p50, pCDNA3.1/p65 and pCDNA3.1/IRF-5 in comparison to the dual cotransfection with pCDNA3.1/p65 and pCDNA3.1/IRF-5 only.

Taken together, p50, cRel and IRF-5 induce the basal activity of the MP -261 approximately 5 fold in comparison to the transfection with the empty vector backbone. This induction is increased to approximately 10 fold by the cotransfection with the dual combination of pCDNA3.1/IRF-5 and pCDNA3.1/cRel, but is not altered by the combination of pCDNA3.1/p50 and pCDNA3.1/cRel. A much stronger induction of approximately 33 fold is achieved by the cotransfection with pCDNA3.1/p65 that is significantly enhanced to approximately 45 fold by the dual combination of pCDNA3.1/p65 and pCDNA3.1/IRF-5. The induction by the pCDNA3.1/p65 alone or by the combination of pCDNA3.1/p65 with pCDNA3.1/IRF-5 is significantly reduced by the additional cotransfection with pCDNA3.1/p50. These results are in accordance with the data obtained from EMSAs, as p50 and cRel are revealed to bind to the respective NFκB-sites in the MP -261 (see 5.4.).

5.8.3. The combination of p65 and IRF-5 induces the UpP in 293T cells

Although EMSAs did not reveal the binding of NFκB subunits to the predicted NFκB-sites 1 and 2 in the UpP, both sites bound a yet unknown nuclear protein. Furthermore, it has been shown in previous experiments that the UpP functionally cooperated with the MP -261 and the *185 bp enhancer* (see 5.7.1. and 5.8.). As EMSAs revealed the binding of IRF-5 to the IRF-site 3 in the UpP, the previously obtained data was validated in a functional assay showing cooperative binding of the NFκB-subunits and IRF-5. Therefore, 293T cells were cotransfected with pGL3/UpP (Fig. 5.28. A) or pGL3/UpP 3.IRFmut (Fig. 5.28. B) reporter plasmids and different combinations of pCDNA3.1 expression vector coding for p50, p65, cRel, IRF-5 or empty pCDNA3.1 vector as described in chapter 5.8.2. The induction of the UpP or the mutant counterpart by the cotransfection of different combinations of p50, p65 and IRF-5 was compared to the cotransfection with empty vector backbone.

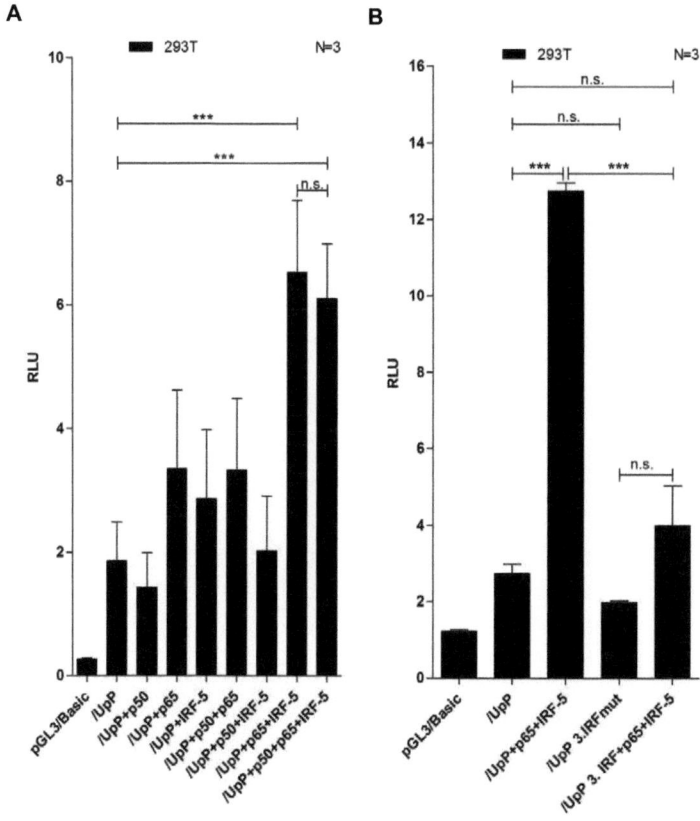

Figure 5.28. p65 and IRF-5 induce the UpP in 293T cells. (A) 293T cells were cotransfected with 0.05 µg of pGL3/UpP reporter plasmid and 0.15 µg of each plasmid coding for the indicated transcription factors (p50, p65 and IRF-5 in pCDNA3.1) or empty vector backbone as control. The total amount of DNA was adjusted with pCDNA3.1 vector backbone to 0.5 µg. The promoter activity was determined by luciferase reporter assays 48 h after transfection. As controls the pGL3/Basic and the pGL3/CMV/luc vector were transfected to assess background activity and for internal normalization, respectively. The resulting RLU were normalized to the protein concentration of each lysate. The cotransfections were compared to the pGL3/UpP control and the resulting p-values were determined via one way ANOVA and Bonferroni's Multiple Comparison *post hoc* test. * $p < 0.05$, ** $p < 0.01$, *** $p < 0.001$ and n.s. not significant ($p > 0.05$). Results represent the means (±SEM) of three independently performed experiments. **(B) The induction of the UpP by p65 and IRF-5 is abrogated when the IRF-site 3 is mutated.** Cells were transfected as described in (A) either with 0.05 µg pGL3/UpP or pGL3/UpP 3.IRFmut reporter plasmid alone or cotransfected with 0.15 µg of each indicated transcription factor (p65 and IRF-5 in pCDNA3.1). The total amount of DNA was adjusted with pCDNA3.1 vector backbone to 0.5 µg. The pGL3/UpP and pGL3/UpP 3.IRFmut setups were compared and the resulting p-values were determined via one way ANOVA and Bonferroni's Multiple Comparison *post hoc* test. * $p < 0.05$, ** $p < 0.01$, *** $p < 0.001$ and n.s. not significant ($p > 0.05$). Results represent the means (±SEM) of three independently performed experiments.

As shown in figure 5.28. (A), the basal activity of the UpP in 293T cells amounted to 1.9 RLU. However, neither the cotransfection with pCDNA3.1 expression vectors coding for p50, p65, IRF-5, the dual combination of pCDNA3.1/p50 and pCDNA3.1/p65 or pCDNA3.1/p50 and pCDNA3.1/IRF-5 induced the UpP (1.4 RLU, 3.3 RLU, 2.9 RLU, 3.3 RLU, 2.0 RLU, respectively). Interestingly, only the combination of plasmids coding for p65 and IRF-5 induced the basal activity of the UpP significantly to 6.5 RLU. This induction was not altered by the triple cotransfection with plasmids coding for p50, p65 and IRF-5 (6.1 RLU). Additionally, a reporter plasmid containing the mutated IRF-site 3 in the UpP was compared to the non mutated UpP to verify the induction of the UpP by p65 and IRF-5 (Fig. 5.28. B). The basal activities of both the mutated and non mutated UpP were not over background (2.0 RLU and 2.7 RLU, respectively), whereas the cotransfection with p65 and IRF-5 induced significantly the non mutated UpP to 12.7 RLU. Importantly, the UpP with the mutated IRF-site 3 was not induced by the dual cotransfection with pCDNA3.1/p65 and pCDNA3.1/IRF-5.

In summary, the basal activity of the UpP is significantly induced by the cotransfection with pCDNA3.1/p65 and pCDNA3.1/IRF-5. The specificity of this effect is maintained by the fact that the UpP containing a mutated IRF-site 3 is not induced by p65 and IRF-5. Furthermore, these results are in accordance with those from ESMA analyses, as IRF-5 and an unidentified factor have been shown to bind to the respective IRF- and NFκB-sites in the UpP (see 5.4.).

5.9. The transcription factors IRF-1, IRF-2, IRF-5, p50, p65 and cRel are differentially expressed in iDCs, mDCs and HFF cells

To further support the involvement of the NFκB- and IRF-families as regulators of CD83 expression, the nuclear expression patterns of the transcription factors IRF-1, IRF-2, IRF-5, p50, p65 and cRel in monocyte derived iDCs, monocyte derived mDCs and HFF cells were examined. Therefore Western blot analyses were performed using nuclear extracts from these three different cell types (Fig. 5.29. A). Furthermore, results obtained from Western blot analyses were evaluated statistically by quantifying the signal intensities of the bands using the AIDA software (Fig. 5.29. B).

Results

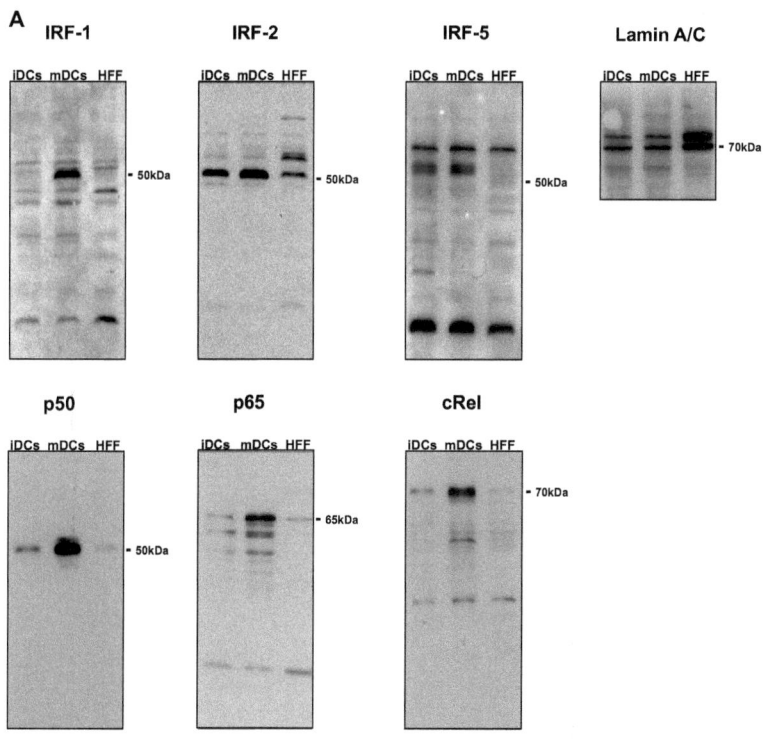
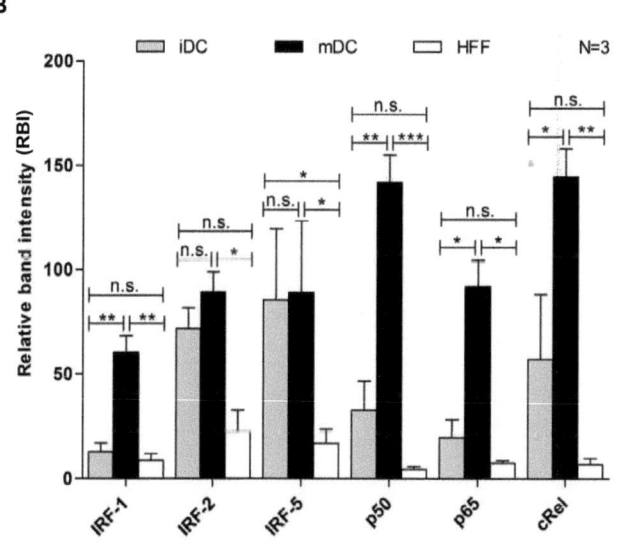

Results

Figure 5.29. IRF-1, IRF-2, IRF-5, p50, p65 and cRel are differentially expressed in iDCs, mDCs and HFF cells. (A) For Western blot analyses nuclear extracts derived from monocyte derived iDCs, 20 h LPS-matured monocyte derived mDCs or HFF cells were used. The nitrocellulose membranes were stained with rabbit anti-human p50, p65, cRel, IRF-1, IRF-2, goat anti-human IRF-5 and Lamin A/C antibodies (loading control). The signal was detected via chemiluminescence. One representative experiment out of three independently performed experiments is shown. **(B)** The band intensities were quantified and normalized to the Lamin C (70kDa) loading control using the AIDA software. The band intensities of iDCs, mDCs and HFF cells were compared for each protein (p50, p65,cRel, IRF-1, IRF-2, and IRF-5) and the resulting p-values were determined via one way ANOVA and Bonferroni's Multiple Comparison *post hoc* test. * $p < 0.05$, ** $p < 0.01$, *** $p < 0.001$ and n.s. not significant ($p>0.05$). Results represent the means (±SEM) of three independently performed experiments.

As shown in figure 5.29. (A), the Lamin loading control (~70 kDa) showed that equal amounts of nuclear extracts were loaded for iDCs and mDCs, whereas a slight excess was loaded for HFF cells. IRF-1 is only expressed in the nucleus of mDCs indicated by a band at ~50 kDa. IRF-2 (~50 kDa) is expressed in all three cell types, although there is a stronger expression in iDCs and mDCs in comparison to HFF cells. IRF-5 (~50 kDa) is not expressed in the nucleus of HFF cells, but in the nucleus of iDCs and mDCs. The NFκB-family members p50 (~50 kDa), p65 (~65 kDa) and cRel (~72 kDa) are only present to a low extent in the nucleus of HFF cells and iDCs, but to a much stronger one in mDCs.

Figure 5.29. (B) shows the statistical evaluation of the Western blot data: IRF-1, p50, p65 and cRel are significantly stronger expressed in the nucleus of mDCs than in iDCs and HFF cells. IRF-2 and IRF-5 are equally expressed in the nuclei of iDCs and mDCs, but weaker (only statistically significant for IRF-5) in HFF cells.

In summary, the nuclear expression observed by Western blot analyses and the respective statistical evaluation for three different donors are in accordance with the results obtained from EMSAs as well as from cotransfection experiments. Moreover, it suggests that the induction of the CD83 expression requires factors that are cell type- and status-specific in their own right.

Results

5.10. The DNA in the promoter region of CD83 is not differentially CpG methylated

To elucidate, whether CpG methylation plays a role in the regulation of the CD83 promoter, pyrosequencing analyses of the CpG dinucleotides in the area comprising the nucleotides bp -502 to bp +212 (CD83 transcription start at bp +1) were performed in cooperation with *Varionostic* (Ulm). This 714 bp long area of the CD83 gene locus, comprising the UpP and the MP -261, displays an above average density of CpG dinucleotides. As highlighted in figure 5.30., especially the area of the 97 CpG dinucleotides located in between the arrows was of great interest, as general criteria for a CpG island are fulfilled: (i) A GC content of over 55% (74.1%), (ii) an observed versus excepted CpG ratio above 0.6 (1.05) and (iii) a length of over 200 bp (673bp). Of particular importance are especially those CpG dinucleotides that lie within an IRF or NFκB binding site of the proposed ternary complex of UpP, MP -261 and *185 bp enhancer*. In order to assess cell type specific epigenetic regulation, the CpG methylation level of iDCs, mDCs and HFF cells have been analyzed. Therefore, monocyte derived iDCs were either matured for 20 h with LPS or left untreated and harvested along with the HFF cells. Next, total genomic DNA was prepared from the three cell types and DNA samples were analyzed by *Varionostic* (Ulm).

Results

```
TCACAGTATTGTTTCTAATATTAGTAATACAAAAAGAAACAACCTGTATGTCTAACACTA
ATCGATTCTAATTTATGGTGCAACTGAACAATGGACCAAAATGATGCTGTTGGAAGTTT
TTAATGATGTGGAACCGCTTGCAAATTATTAAGCTAAAAGAAAGTAGGTTACAAGATAG
CAGGAAGAATAAACCATTAAAAATACCAATCTGTGCACTGACAAATGTTATAAATATTTT          CD83 upstream
ACGTTATGTTATGTTATAAACATTTTATAATATAAAAAAATGTTAACTGAAGTTACTTCCT         sequence
GGATGAATTACAGGTGATTTCATTGTCTTCTAGAATTTTCTTTTCCAAAAATGTTGTGTA
TGCCGTGTAATTATTATTTTAATAGGAGACACTCTCCTTTGGTGATATAATTTAAACAGGA
CGGTACTGACTGATAACCTCCCGGGGAAGGCAGGGAGCCAAGTACTACAGACTTGTA
TGTTTCCATGGAAATCTAACGCGCCTTTGATTATCACAGATTCTGGAGAAGAGTGAGG
ACTTGGGTTCACCAGT
                                                  GCGTTCCC            Upstream
AAGGACAGGCTGGGCTTCTGAGGAAGTTGCCCACCC━━TCGGAATCTGGTTTGGCCT           promoter (UpP)
CCGTAAAATGGGCAGATCCCGCTCGGATGGCCCGGTTCCCGGCTTCCTTTTGCGGGT
CAACGGCAGCGTCACGCGCGCGAGCGCGGTCTGCAAAGCCCCCA

GCGCTGGGCGTCACGCGGGGATTGCTGTCGCCGCTGCCAGCCGCAGCAGCGACGC          Intermediate
GAACTCGGGGCGCCCGGCCCGGGCGCGC                                        sequence

GGGGGCGGGGACGCGCACGCGGCGAGGGCGGCGGGTGCGACGGGGGCGGGGA
CGGGGGCGGGGACGGGGGCGAAGGGGGCGGGGACGGGGGCGCCCGGCCTAAG             Minimalpromoter
CGGGACTAGGAGGGCGCGCCACCCGCTTCCGCTGCCCGCCGGGGAATCCCCGGG              (MP -261)
CTGGCGCGCAGGGAAGTTCCCGAACGCGCGGGCATAAAAGGGCAGCCGGCGCCG
CGCGCCACAGCTCTGCAGCTCGTGGCAGCGGCGCAGCGCTCCAGCC

ATGTCGCGCGGCCTCCAGCTTCTGCTCCTGAGCTGCGGTAGGGCTCGCGAGCGCCT           CD83
GTCTCGCCTGTCGCCCCCCGCCCCTCCACGACACCCCTCCCGTCGGTCGCTTGCTC           Transcriptionstart
ACGACGCGCTCTCTCTTTCTTGTAGCCTACAGCCTGGCTCCCGCGCGACGCCGGAGGTG        and intron 1
AAGGTGGCTTGCTCCCAAGATGTGGACTTGCCCTGCACCGC━━CTGGGATCCGCA
GGTTCCCTACACGGTCTCCTGGGTCAAGGTAGGTGCTGCGATACCCACGGGCTGGG
GTTTGGTGGGCTCATTTGAAGACAGCAGGAACCATCTCCCCTAGGCTGGCGACCCTC
TGTGGCTGCCAGGTGGGGGCGAGGGGCGTCTCCCGCAGCTGAACTTGGAGTACCCA
GCCTCCCGTCGCGCCTCCCCCACCCCATCCGCATCCAGGTACAGGGCCGAATTAGGT
TTTGCTCTCCGCAGACCTCAATCCCCTTCCTGTCACTGAAGGTGGCCTGAGATGAATG
ATCCACTTAAGATGTTTTGGAAGGGCAGAGACTCTCATTTGGATTAATTCTGGAGGCC
ACCTGTGGTTGTGGGCCAGCAGGTCAGGAAGAAAGCAACAGGGACCTAGATTTGGGC
ATTGGACAGGGGGAATGTCTCCAGACTTCTGATTTCTTGTGTTTTGTGACTGTGATGC
CCATGATACATGGGAGGGGGAGGGGGCAATTTGAAAGGAAAGGCTAAGACACAGAA
GTGACTTAGGCCATTTCATCCATGGTAGTTATCAGTGGTCATCTCCTTTGTGGGGATAC
CCTTGGCTTCCTCCCCTAGCCCTCCTCCTCCTTCCTCTGGCAGCCTTGAGAGGCATCAG
GTGGATGCATGAGCCGGAGCCCGCATGTGTGAAGAACAGGCCTTGCTGCTCCTACTGT
AAGTGGACTGAGTGACAAGGAGGCTTTTTCAAGGTTTCCTCTTGACTGAAACATTCTC
AGATTCTAAGATGGCAATGATGGTGTCATTCCAAAGCCAAGCAGCTACTGTTTGATAT
CACTGGTCCTTCTTTAAGTCAGGCCACTGCTACCACAGCACCTCCATTTTAACCCAAAT
GAATATGATATTACAACCTTACTCTGTAGCTCTCACTGATTTGCTGTCTTACCACGGGG
GCAAATCTCTGCACTTGTAGCTTTCCCCAAAATGCAGGGCGTTCTTCTGCCCACCATA
AAAGATACTATAAGAAACTGTACGTCTTTGGCCACTTAACAGTACAAGGCATCATTGCG
GTGATCTCTTTGTGTGTGTGTCTCCTAACTGGATGGTCAGTTCCCTGGGGGGCAGTG
GCTGTATCCATACTTCTGTGTATTCTTCACGGCACCTAATTTTTGCCCTATAAATTGCA
AAGGTGCTCTGTGAATTCAGCCCAGCACTTCATGAGTTATGCATGACGGGGATGGTG
CTGCTGCCTCAGAGCATTGTATTG
```

SP1-sites:	NFκB-sites:
CTGAGGAAGTTGCCC_SP1-site 1	GGGCAGATCCCGC_NFκB-site 1
GGGGCGGGGA_SP1-site 2	TGCAAAGCCCCCA_NFκB-site 2
GCGAGGGCGGCGGGTG_SP1-site 3	CGGGGGCGCCCCG_NFκB-site 3
GGGGCGGGGA_SP1-site 4	GGGAATCCCCCG_NFκB-site 4
GGGGGCGGGGA_SP1-site 5	CAGGGAAGTTCCC_NFκB-site 5
GGGGCGGGGA_SP1-site 6	
GCCTAAGCGGGACTA_SP1-site 7	CG: CpG di-nucleotides

IRF-site:
TTCCCGGCTTCCTTTTGCGGG_IRF-site 3

➡ ⬅ Area of CpG methylation analysis

Figure 5.30. Schematic depiction of the nucleotide sequence of the CD83 promoter and the surrounding area. CpG dinucleotides are marked in grey, SP1-, IRF- and NFκB-binding motifs are bold and underlined. The area comprising the 97 CpG dinucleotides, which were analyzed for CpG methylation, is marked between the two arrows. The sequence upstream of the CD83 promoter, the UpP, the 82 bp intermediate sequence, the MP -261 and the CD83 transcription start/intron 1 are indicated as separate boxes.

Results

Figure 5.31 displays the percentage of CpG methylation of the analyzed CpGs in iDCs, mDCs and HFF cells. Methylation percentages below 7% were considered as background and therefore regarded as not methylated. Except at the positions 8, 21, 25, 40, 59, 64, 65, 96 and 97 none of the analyzed CpGs was methylated over background in any of the three cell types. These nine CpG residues were considered as outliers, as no true differential methylation was observed. Moreover, the positions were thought to be random and the percentages were minimally over background. The CpG positions 66-77 could not be analyzed by pyrosequencing due to the strong formation of secondary structures.

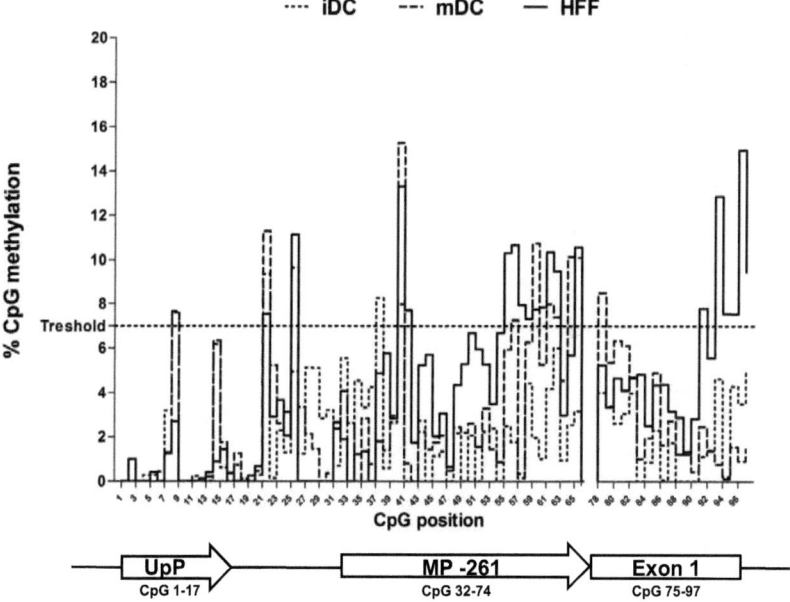

Figure 5.31. The analyzed CpG dinucleotides in the CD83 promoter region are neither differentially methylated in iDCs, mDCs nor HFF. The percentage of methylation for each of the CpG dinucleotides was analyzed by pyrosequencing in monocyte derived iDCs, monocyte derived mDCs and HFF cells. The CpG methylation percentages in iDCs are depicted as a dotted line, in mDCs as a dashed line and in HFF as a solid line. A schematic representation of the CD83 promoter region is aligned beneath the graph. CpGs 1-17 are located within in the UpP, CpGs 18-31 within the intermediate sequence, CpGs 32-74 within the MP -261 and CpGs 76-87 within exon 1 of the CD83 gene. These analyses comprise a gap extending over the CpG dinucleotides 66-77.

Results

In summary no evidence for an epigenetic regulation of the CD83 expression by CpG methylation of the CD83 promoter region was found in iDCs, mDCs or HFF cells. Therefore, epigenetic regulation of the CD83 promoter was excluded, which further supports the biocomputational model of the formation of a tripartite complex of UpP, MP -261 and 185 bp enhancer regulating CD83 expression.

Discussion

6. Discussion

6.1. Full characterization of the human CD83 promoter

Although being expressed in several cell types like activated B and T cells as well as macrophages and epithelial thymus cells, human CD83 is best known as maturation marker for mDCs. DCs strongly upregulate CD83 during maturation within hours after having encountered maturation stimuli such as LPS or TNF-α[46;133]. CD83 is expressed in two isoforms, a soluble and a membrane bound form, which have been shown to be important immune modulators. Studies of several groups demonstrated the crucial role of the membrane bound CD83 in T cell stimulation and in the context of the immune activation in general[127;149;150;426]. Soluble CD83 on the other hand has been shown to be a potent immune inhibitor with great potential for therapeutic applications, such as the prevention of transplant rejection or the treatment of autoimmune diseases like multiple sclerosis[133;427;428]. The diverse functional properties as well as the distinct tempo-spatial expression of CD83 require a tight transcriptional regulation, implicating a complex regulatory mechanism. Berchtold et al. published the sequence of the CD83 minimal promoter (MP - 261) in 2002[167]. However, despite being active in luciferase assays, the MP - 261 showed no cell type specificity. Consequently, additional regulatory elements, such as cell type- and maturation status-specific genetic enhancers, had to be postulated for the tight regulation of CD83 expression in mDCs.

6.2. A 185 bp sequence within CD83 intron 2 has mDC-specific enhancer function

In order to identify potential genetic enhancer regions that regulate mDC-specific CD83 expression, first the H3K9 acetylation level at the CD83 gene locus was examined in iDCs, mDCs and HFF cells using a ChIP-chip™ microarray. It could thereby be shown that a 6 kb region in CD83 intron 2 was specifically hyperacetylated in mDCs, but not in iDCs and HFF cells (see chapter 5.1.). In accordance with the literature it was concluded for several

Discussion

reasons that potential regulatory elements might lay in this region: (i) H3K9 acetylation is associated with open chromatin conformation and thus points towards a high transcriptional activity in this region[227;229]. As this acetylation was highly cell type-specific, one could speculate that the mDC-specific regulation of CD83 expression is controlled by this region. (ii) Eukaryotic gene expression is not only regulated by the genomic DNA sequence, but also by a higher order structure of the chromatin [233;259;429-431]. Many epigenetic modifications, such as histone methylation and acetylation, occur in active chromatin regions at several regulatory elements modifying the chromatin structure. Thereby, especially acetylation of histone lysine residues of promoter regions and enhancers is associated with activated transcription[214;432-435]. In addition, it is known that many regulatory elements carry a cell type- or tissue-specific histone acetylation in a well-conserved manner. Conserved histone H3 acetylation has been shown to occur at several gene loci, such as at the locus control region (LCR) DNase I hypersensitive site elements of avian and mammalian β-globin loci[436-438] or the Eμ heavy chain intron enhancer at the human immunoglobulin heavy chain (IgH) locus[439]. In this respect, it has been shown that histone acetylation is indeed a hallmark of regulatory elements such as genetic enhancers[440]. (iii) CD83 expression is severely affected in human monocyte derived DCs when altering the histone acetylation by treatment with histone deacetylase (HDAC) inhibitors, thus linking CD83 expression to histone acetylation[441].

Indeed, by deletion mutagenesis and luciferase reporter assays a 185 bp fragment within the hyperacetylated region of CD83 intron 2 was identified as a cell type- and maturation status-specific genetic enhancer (see chapter 5.2.). The cell type-specificity of this *185 bp enhancer* was evident in the DC-like cell line XS52, but absent in control cells (i.e. murine NIH3T3 cells, human HeLa cells). Moreover, the *185 bp enhancer* significantly induced the MP -261 in LPS-matured DCs, whereas in iDCs the induction was not significant. The slight induction in iDCs might be due to the physical stress caused by the electroporation procedure inducing maturation of iDCs and resulting in slight activation and in a weak induction of the MP -261 by the *185 bp enhancer*. This problem was addressed in the following experiments, where for the transduction of iDCs adenoviral vectors instead of electroporation were used (see chapter

Discussion

6.4.). Consequently, the maturation status of the iDCs was not altered by the adenoviral transduction.

Interestingly, no induction of the MP -261 by the *185 bp enhancer* was observed in Raji B cell and the Jurkat T cell lines, although both express CD83 on their surface. This suggests a completely different regulation mechanism of CD83 expression in B and T cells. Taken together, in human monocyte derived DCs the *185 bp enhancer* displayed not only cell type specificity, but also maturation status specificity.

6.3. A ternary complex consisting of MP -261, *185 bp enhancer* and UpP regulates mDC-specific CD83 expression

The functionality of the *185 bp enhancer* was validated by luciferase assays, as the *185 bp enhancer* strongly induced the CD83 minimal promoter in a cell type- and maturation status-specific manner in XS52 cells and mDCs. To assay the underlying molecular mechanism, especially the transcription factors binding to the promoter and the *185 bp enhancer*, a bioinformatical analysis was performed in collaboration with Dr. Thomas Werner (Munich). The biocomputational model developed by Dr. Werner predicted that in human mDCs the CD83 gene expression is driven by a bipartite promoter region containing two cassettes of transcription factor binding sites (TFBS). Three NFκB-sites in the proximal promoter region (MP -261; NFκB sites 3-5) and one IRF-site flanked by two additional NFκB-sites in the distal promoter region (IRF-site 3 and NFκB-sites 1 and 2), termed the CD83 upstream promoter (UpP). These transcription factor (TF) cassettes are proposed to cooperate with the *185 bp enhancer* located approximately 5 kb downstream of the promoter within the second intron of the CD83 gene, which contains two IRF-sites (IRF-site 1 and 2). An overview of all biocomputationally predicted TFBS for the CD83 regulatory elements is shown in figure 6.1.

Discussion

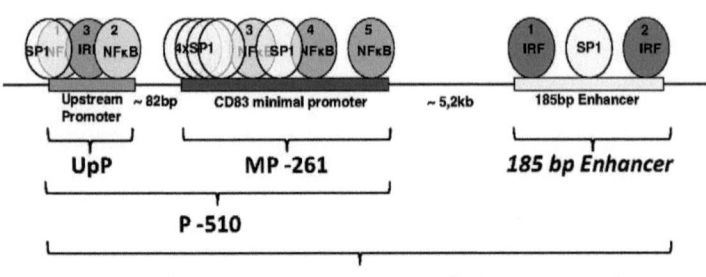

DC-specific tripartite CD83 promoter/enhancer complex

Figure 6.1. Schematic representation of the predicted transcription factor binding sites in UpP, MP -261 and 185 bp enhancer.

The model proposed that all three regions, UpP, MP -261 and 185 bp enhancer, are essential for maximal promoter activity, working synergistically together. Thereby, NFκB binds directly to the three NFκB-sites in the proximal promoter. As the bioinformatical model predicted the two NFκB-sites in the UpP as weak binding sites, they might only be engaged cooperatively by IRF- and NFκB- factors. Notably, one of the NFκB-sites in the MP -261 has already been experimentally verified and published by Berchtold and colleagues[167].

Furthermore, the model proposed that all three regions join into one ternary complex. This is mediated by NFκB-IRF protein interactions by the folding of the distal promoter (UpP) onto the proximal promoter (MP -261) and the 185 bp enhancer looping into this complex. This tripartite alignment generates three copies of a well-known NFκB-IRF transcriptional module (Fig. 6.2.).

Discussion

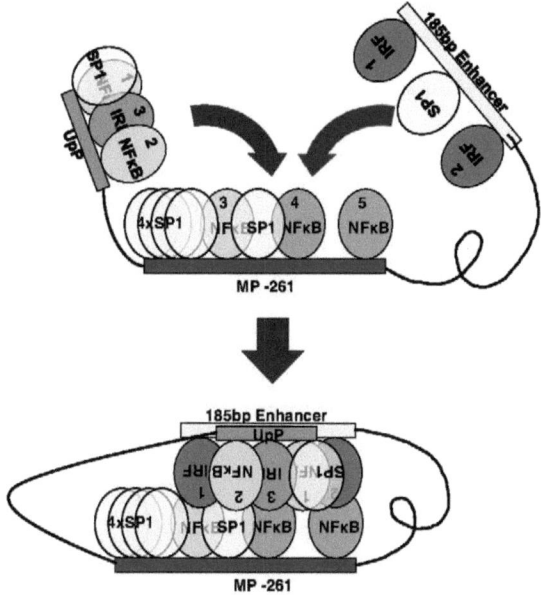

Figure 6.2. Schematic representation of the tripartite complex consisting of UpP, MP -261 and 185 bp enhancer.

The physical interaction of members of the NFκB- and IRF-transcription factor families (e.g. p65 with IRF-1 or IRF-5)[425;442] followed by the activation of gene expression is a common mechanism and has been described for e.g. IL-12p35, IL-15, type I interferons, VCAM-1 and MHC class I[442-446]. Remarkably, in the case of CD83 the TFBSs composing these transcriptional modules are not located in cis, but form in trans by folding all three regions into one DNA-protein complex. Furthermore, the relative distances between the distal and the proximal promoter provide an accurate "key to lock" principle as the folding of the distal part onto the MP -261 leads to a close to perfect alignment of the TFBSs in both regions with one IRF-NFκB module in the middle flanked by two NFκB-site pairs. The IRF-sites in the enhancer complement the flanking NFκB-site pairs to form two additional NFκB-IRF modules.

In this context, the biocomputational model also predicted five SP1-sites in the CD83 minimal promoter as well as one SP1-site in the 185 bp enhancer. The four distal SP1-sites have already been experimentally verified as coinducers of

Discussion

the MP -261 by Berchtold and colleagues[167]. However, the relevance of the two remaining potential SP1-sites remains to be elucidated. For the predicted ternary complex the SP1-sites are thought to have only a supportive role and/or to act as stabilizing elements for the formation of the tripartite promoter/enhancer complex. Both the proximal promoter structure of the three NFκB-sites as well as the two IRF-sites in the enhancer are phylogenetically well conserved in at least six different species (human, mouse, rat, dog, cattle, and pig), while the distal promoter (UpP), containing two NFκB-sites and one IRF-site, appears to be present only in humans.

The formation of the bioinformatically predicted tripartite promoter/enhancer complex can only be proven accurately with adenoviral reporter constructs. The large adenoviral vectors favor the intramolecular interaction of the promoter and enhancer elements, whereas small plasmids interact rather on an intermolecular level. Using the pGL3 luciferase reporter plasmids, the induction of the MP -261 was most likely not facilitated through intramolecular (*cis*) looping of the enhancer towards the promoter, but rather by means of intermolecular (*trans*) interactions of one pGL3 plasmid providing the MP -261 and one or more different plasmids providing the enhancer sequence. Therefore, several adenoviral constructs were generated, containing different combinations of UpP, Mp -261 and *185 bp enhancer* (see chapter 5.4.). This allowed the examination of the interaction of UpP, MP -261 and 185 bp enhancer in a more native, chromosome-like conformation.

Using adenoviral vectors, only constructs bearing all three regulatory elements showed an induction of the luciferase expression in LPS-matured DCs, but not in iDCs. Vectors lacking either the UpP, the *185 bp enhancer* or both elements, showed only basal transcriptional activity, independent of LPS-stimulation. These results strongly support the biocomputational model, forming a tripartite regulatory complex in mDCs.

Interestingly, the MP -261 showed only a basal transcriptional activity in Raji B cells, Jurkat or JCAM T cells with adenoviral luciferase transporter constructs, which was not influenced by the UpP and the *185 bp enhancer*. This is in accordance with the previous data derived from the electroporation experiments. However, there is a basal transcriptional activity of MP -261 in these cell lines, which can be slightly enhanced by LPS or TNF-α/PGE$_2$

Discussion

stimulation, respectively. These inflammatory stimuli are well known to induce NFκB-activation in B and T cells[447;448], which in turn might lead to an induction of the MP -261, regardless of the presence of the UpP and/or the *185 bp enhancer*. Accordingly, Raji B cells and both T cell lines upregulate CD83 expression on their surface upon LPS and TNF-α/PGE$_2$ stimulation, respectively. However, the precise regulation of CD83 expression in B and T cells has not been studied yet.

Taken together, it could be confirmed that UpP, MP -261 and *185 bp enhancer* cooperate to form a tripartite regulatory complex in mDCs and that all three elements are indispensable for the formation of this complex. In Raji B and Jurkat and JCAM T cells the MP -261 is not induced by the UpP and the *185 bp enhancer*. Although all three cell lines express CD83, the formation of a ternary complex composed of UpP, MP -261 and *185 bp enhancer* might not play a role in these cells. Therefore, it was concluded that the regulatory mechanisms of CD83 expression in these cells differ completely from DCs.

6.4. The formation of the ternary complex is mediated by NFκB- and IRF-transcription factors

The functionality and the cooperation of the individual elements UpP, MP -261 and *185 bp enhancer* has been demonstrated in human mDCs. The next step was to investigate the role of the individual TFBSs. Therefore, each predicted NFκB- and IRF-site was first analyzed for binding of nuclear factors in iDCs, mDCs and HFF cells using EMSA (see chapter 5.6.). Furthermore, the functionality of the NFκB- and IRF-sites was assessed with "*gain of function*" and "*loss of function experiments*", respectively (see chapters 5.7. and 5.8.). Finally, the expression patterns of both NFκB- and IRF-transcription factors were assessed by Western blot analyses in iDCs, mDCs and HFF cells (see chapter 5.9.).

Discussion

6.4.1. Verification of the NFκB-transcription factor binding sites

EMSA clearly revealed the binding of the nuclear NFκB-subunit p50 to the predicted NFκB-sites 3, 4 and 5 in the MP -261 in both iDCs and LPS-matured DCs by supershifts. Moreover, cRel was observed to bind to the NFκB-sites 4 and 5 only in mDCs, Furthermore, binding of a yet unidentified factor could be demonstrated for the predicted NFκB-sites 1 and 2 in the UpP. However, in accordance with the bioinformatical model all five predicted NFκB-sites were verified as being capable to bind nuclear proteins. The finding that p50 and cRel might be involved in the regulation of CD83 expression in mDCs is supported by several reports from other groups. NFκB-subunit p50 confers the binding domain in both p50/p65 and p50/cRel heterodimers and is therefore involved in most steps of NFκB-mediated DC development and function[308;313;347;383]. Accordingly, the binding of p50 to the predicted NFκB-sites 3, 4 and 5 in the MP -261 was expected.

Furthermore, these data are in accordance with the EMSA results for NFκB-site 4 published by Berchtold and colleagues[167]. Additionally, the binding activity is increased in mDCs for the NFκB-sites 3, 4 and 5 in comparison to iDCs, indicating also a maturation specific function. Indeed, several NFκB-subunits play an important role for maturation-specific events in the DCs[383]. Interestingly, no binding of p65 to any of the oligonucleotides coding for the predicted NFκB-sites was observed in EMSA, which is surprising for several reasons: First, p50 and p65 form the most abundant and prototypical transcription-activating heterodimer of NFκB and are an integral part of TLR-mediated signaling[280;308;309;347;383;449]. Second, p65 was reported to bind to the NFκB-site 4 in EMSAs performed by Berchtold et al., although this group did not use nuclear extracts from primary cells, but whole cell lysates from the murine DC-like cell line DC 2.4[167]. Third, p65 strongly induced the MP -261 in cotransfection experiments performed in 293T cells, further supporting the results from Berchtold et al. This suggests a binding of p65 to at least one of the predicted NFκB-sites and therefore its involvement in CD83 regulation. On the other hand some aspects contradict the involvement of p65 in the cell type- and maturation-specific regulation of CD83 in DCs: Although the nuclear translocation of p65 in

Discussion

human DCs is increased upon LPS-stimulation[450], it is not influenced by stimulation with a maturation cocktail composed of IL-1α, TNF-α, IL-1β, IL-6 and PGE_2[343]. In contrast, CD83 is strongly upregulated upon both maturation stimuli, which suggests an additional factor than p65 involved in CD83 regulation.

It has been reported that p50, p65 and cRel can compensate each other's function as demonstrated by p50, p65 and cRel knockout mice. The knockout of a single NFκB-subunit did not impair the DC development and function, whereas the double knockouts of either p50/p65 or p50/cRel severely affected different aspects of DC biology like development and survival[451]. In this context, it is noteworthy that also cRel significantly induced the MP -261 in cotransfection experiments, albeit to a lesser degree than p65. Unlike p65, NFκB-subunit cRel is closely connected to mDCs: NFκB-subunit cRel specifically accumulates in the nucleus of human and murine mDCs, whereas it is almost absent in iDCs[451;452]. This is further supported by the fact that LPS-maturation induced especially p50/cRel containing NFκB-complexes in mice whereas p65, albeit induced to some extent, was significantly less present[451]. Furthermore, the p50/p65 heterodimer was shown to exert important functions rather in the early stages of the DC lifespan, like development. On the other hand, the maturation induced p50/cRel heterodimers were shown to facilitate essential functions for the already activated DCs, such as survival, cytokine production (IL-4, IL-12, IL-23 and IFN-γ) and T cell stimulatory capacity[343;443;451-455].

However, functional evidence for the NFκB-sites had still to be generated by cotransfection experiments for two reasons: (i) For the predicted NFκB-sites 1 and 2 in the UpP EMSA results were ambiguous. (ii) EMSA is an *in vitro* approach with isolated TFBSs taken out of the cellular context with no possibility for cooperative binding or epigenetic regulation like e.g. histone modifications. Therefore, engagement of otherwise potential *in vivo* binding factors might be hampered by missing epigenetic factors, the lack of neighboring and/or cooperative binding sites and displacement by nuclear factors. In natural circumstances, these factors would be restricted in accessing the binding sites, but are freely floating in the nuclear lysates used for these assays.

Discussion

The combination of cRel and IRF-5 induced the MP -261 significantly stronger than cRel or IRF-5 alone. No increased induction was achieved, when p50 was cotransfected with cRel or IRF-5, when compared with the single transfection of these factors. The induction of the MP -261 by p50 and cRel was in accordance with the results derived from EMSA, as both factors bound to the NFκB-site 4 and 5 and p50 bound alone to the NFκB-site 3. In this context, p50 might interact with a yet unknown factor expressed in DCs to induce the MP -261. The induction by IRF-5 on the other hand was surprising, as the MP -261 does not contain an IRF-site. Interestingly, IRF-1 and IRF-2 did not show any effect on the MP -261 (data not shown). However, the recruitment of IRF-5 to a NFκB-site mediated by p65 has been described before by Krausgruber and colleagues[425]. They propose a model where p65 alone induces only basal transient transcriptional activity of the TNF-α promoter in DCs. Upon LPS-stimulation IRF-5 is recruited on the one hand to an upstream ISRE element flanked by an NFκB-site and on the other hand interacts at another downstream NFκB-site directly with p65 resulting in an upregulated and prolonged induction of the TNF-α expression[425]. The model of the TNF-α promoter regulation by an upstream and a downstream regulatory element involving IRF-5 upon LPS-stimulation shows striking resemblance to the induction of the MP -261. A basal transient transcriptional activity of the MP -261 may be achieved by low levels of c-Rel and/or p65 in iDCs and other cell types, respectively. Upon stimulation of iDCs by e.g. LPS, IRF-5 is recruited to the MP -261, interacts with p65 or cRel and then mediates prolonged and increased CD83 expression, as only in the presence of these factors the ternary complex can form in mDCs.

Interestingly, the cotransfection of p65 and the MP -261 luciferase reporter plasmid induced the MP -261 much stronger than cRel, IRF-5, p50 or the combination of cRel and IRF-5. This induction could be significantly enhanced further by the cotransfection of p65 and IRF-5, but not by the combination of p65 and p50. Cotransfection of p50 and p65 or p50, p65 and IRF-5 even reduced the induction of the MP -261 as compared to p65 alone or the combination of p65 and IRF-5 without p50. A possible explanation could be that due to the *in vitro* overexpression of p50 and p65 also p50 homodimers may form, which display a higher DNA binding affinity than the p50/p65 heterodimers. These p50 homodimers then displace the p50/p65 heterodimers

Discussion

and induce the MP -261 to a much lesser extent than p65 in combination with endogenous p50[456]. However, p65 was not detected in EMSA to bind to any predicted NFκB-site. This might suggest that CD83 regulation *in vivo* is rather mediated by cRel adopting the role of p65 in DCs as opposed to the previously described TNF-α model. The interaction of cRel and IRF-5 has yet to be assessed.

Another possibility might be that p65, cRel and IRF-5 coregulate CD83 expression. For the basal or transient transcription, p65 might be the responsible factor. Maturation- and cell type-specific CD83 expression could then be achieved by cRel and IRF-5 coregulation. Very strong evidence for the functionality of the NFκB-sites 1 and 2 and the IRF-site 3 in the UpP arose from further cotransfection experiments in 293T cells, where the UpP alone was induced by the specific combination of p65 and IRF-5 in luciferase reporter assays. NFκB-family member cRel failed to induce the UpP, regardless of the cotransfection with IRF-5 (data not shown). As EMSA did only reveal the binding of an unknown factor to the predicted NFκB-sites 1 and 2, p65 might as well be a candidate to bind cooperatively with IRF-5 to the NFκB-IRF-NFκB-module in the UpP. This assumption is in accordance with the results from EMSA showing a binding of IRF-5 to the UpP. These results further support the involvement of p65 in the regulation of CD83 in DCs, but do not rule out cRel as transcription factor binding to the MP -261.

Furthermore, a strongly upregulated nuclear translocation of p50, p65 and cRel upon LPS-maturation could be shown using Western blot analyses with nuclear extracts generated from iDCs and mDCs. These results support the previous findings and explain the stronger binding of nuclear proteins to the NFκB-sites 2, 3, 4 and 5 in EMSA. Therefore, it was concluded that all three NFκB-family members are specifically activated in LPS-stimulated human monocyte derived DCs and translocate to the nucleus. These results further support that the cell type- and maturation -status-specific formation of the ternary complex regulating CD83 expression is mediated by maturation status-specific NFκB-factors and thereby further underline the functionality of the predicted NFκB-sites.

Discussion

6.4.2. Verification of the IRF-transcription factor binding sites

For the predicted IRF-sites, supershift experiments revealed the binding of IRF-1 and IRF-2 to the predicted IRF-site 1 in the *185 bp enhancer* in both iDCs and mDCs. As for IRF-site 2 in the *185 bp enhancer*, strong binding of a yet unknown nuclear protein was observed (iDCs and mDCs). Furthermore, the binding of IRF-5 to the predicted IRF-site 3 in the upstream promoter was detected in mDCs. Most importantly, all three predicted IRF-sites were functionally verified in EMSA, further supporting the biocomputational model of the ternary complex regulating CD83.

According to the literature the binding of IRF-1 and IRF-2 to the predicted IRF-site 1 in iDCs and mDCs suggest a common principle of IRF-mediated transcriptional regulation in DCs. Both IRF-family members are considered to be essential mediators of DC development and function[354;355] and are able to bind to the same TFBS[349;377]. This explains why in EMSA, both factors were identified to bind to the same IRF-site. Typically, they regulate gene expression in an antagonistic manner by competing for the same TFBS, with IRF-1 as transcriptional activator and IRF-2 as repressor[367;377;415;457;458]. In some cases, like the regulation of IL-12p40 or Cox-2, IRF-1 and IRF-2 cooperate as activators[351;459;460]. Thus, both factors may be part of an elaborate mechanism for the fine-tuned regulation of CD83. Indeed, in iDCs the signal intensity of the IRF-1 supershift was lower than the supershift intensity of IRF-2. Conversely, in mDCs the supershift intensity of IRF-1 raised and the IRF-2 supershift intensity decreased to a point where the IRF-1 supershift intensity exceeded the IRF-2 intensity. These results suggest a regulatory mechanism mediated by IRF-1 and IRF-2, thereby competing for the same TFBS. Accordingly, the group of Lionel B. Ivashkiv reported the upregulation of nuclear expression of IRF-1 in human DCs mediated by maturation stimuli such as LPS and TNF-α/PGE$_2$[461]. It is conceivable that IRF-2 acts in this context as a transcriptional repressor, which is then displaced by the strongly upregulated IRF-1. This assumption is supported by findings from the groups of Maniatis, Wathelet and Sgarbanti, which describe IRF-1 as enhancer binding factor that activates IFN type I, but also HIV gene transcription, respectively[260;446;462-464]. In accordance with the previously described cell type specific H3K9 acetylation of the first 6 kb of CD83

Discussion

intron 2, it is noteworthy that both IRF-1 and IRF-2 interfere with histone acetylation[465]. The repressive function of IRF-2 is partially attributed to its ability to inhibit acetylation of core histones chromatin which leads to a closed conformation[466]. IRF-1 on the other hand is able to recruit histone acetyltransferases (HAT) that in turn acetylate core histones leading to an open conformation of the chromatin, which allows the binding of transcription factors as well as DNA-looping to establish e.g. enhancer/promoter interactions[231;440;465;467].

With regards to the fact that EMSA is an *in vitro* approach with isolated TFBSs taken out of context (see 6.5.1.1.), not addressing epigenetic regulation (e.g. histone modifications), no assertion could be drawn for the predicted IRF-site 2, other than it is able to bind nuclear factors from both iDCs and mDCs.

The predicted IRF-site 3 bound the IRF-family member IRF-5 in both iDCs and mDCs. Although IRF-5 is expressed constitutively in DCs[425], human IRF-5 molecules are expressed as multiple splice variants with distinct cell type–specific expression, cellular localization, differential regulation/activation and dissimilar functions[382;468;469]. The functions of IRF-5 are not fully elucidated yet, but it is known that IRF-5 is a major mediator of MyD88-dependant TLR-signaling and an essential transcription factor for the expression of TNF-α, IL-6 and IL-12[382;470]. In this context, IRF-5 is an adequate candidate for the fine regulation of CD83 by means of the UpP. IRF-5 is not only linked to TLR signaling, but it has been described recently by the group of Irina Udalova to coinduce gene expression in immune cells as upstream promoter binding factor[425]. In a later publication the same group reported the upregulation of CD83 in human DCs after transduction with an adenoviral vector coding for IRF-5[471]. Furthermore, IRF-5 has been shown to be essential in the induction of CD83 expression, since its cotransfections in 293T cells turned out to be essential in inducing the upstream promoter in cooperation with p65. Importantly, deletion and mutation experiments verified the need of all three predicted IRF-site for the functionality of the *185 bp enhancer* and the UpP in the formation of the ternary complex.

Western blot analyses showed that nuclear expression of IRF-1 is strongly upregulated during DC maturation, whereas the levels of IRF-2 remained constant. This supports the assumption that the nuclear translocation of IRF-1 might displace the constitutively expressed IRF-2 at the *185 bp enhancer* in mDCs leading thereby to an activation of the ternary complex. Accordingly, IRF-1 is, unlike IRF-2, not expressed in HFF cells. Moreover, IRF-5 is expressed in the nucleus of both iDCs and mDCs, but not in HFF cells. Therefore, it is conceivable that the DC-specific factor IRF-5 might in fact strongly contribute to the cell type-specific expression of CD83 by the ternary complex.

Taken together, EMSAs, cotransfection, mutation experiments and Western blot analyses confirmed the bioinformatically predicted NFκB- and IRF-sites as well as their functionality in the formation of the ternary complex. NFκB-family members p50, p65 and cRel cooperate with the IRF-family members IRF-1, IRF-2 and IRF-5 to regulate CD83 expression forming a tripartite promoter/enhancer complex. In this context, cRel and IRF-1 as well as IRF-5 were shown to play a role as maturation-specific inducers of CD83 expression. The cooperation of these factors to regulate immune related gene expression has already been reported before. Several groups demonstrated the direct physical interaction of IRF-1, IRF-2 and IRF-5 with NFκB-family members p50 and p65. Moreover, also the distinct binding and functional cooperation of IRF-1 with p50/p65 as well as of IRF-1 and IRF-5 with cRel in the induction of TNF-α, IL-12 and GBP-1, respectively, has been shown to date [425;443;472]. Thus, the involvement and cooperation of these factors is well known and therefore very plausible also for CD83 gene regulation. In addition, the striking novelty of this thesis work is the complex tripartite formation of the regulatory IRF-NFκB-modules *in trans* involving all described transcription factors simultaneously.

6.5. Summary and future prospects

Taken together, the cooperation in one complex of three distinct CD83 promoter elements, namely the UpP, the MP -261 and the *185 bp enhancer* was demonstrated. Thereby, the interaction of the DNA elements is mediated by IRF- and NFκB-factors. DC-specific factors like cRel and IRF-5, the maturation status-dependent upregulation of IRF-1 and more generic expressed factors like IRF-2 and p65 cooperate in a complex system to regulate cell type- and maturation status-specific CD83 expression. A model for the transcriptional regulation of CD83 is shown in figure 6.3.

Discussion

A

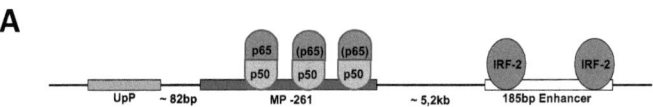

- Binding of NFκB p50/p65 to the CD83 minimal promoter
- Inhibition of the *185 bp enhancer* by constitutive binding of IRF-2

Basal/transient transcription in iDCs and non-DC cell types

B

↑ TLR-induced upregulation/activation of IRF-1, IRF-5 and cRel during DC maturation

C

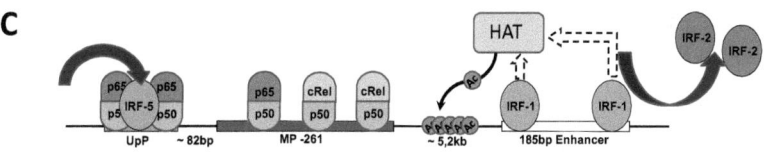

- Displacement of repressing IRF-2 by activating IRF-1
- Acetylation of histone 3
- Cooperative binding of NFκB p50/p65 and IRF-5 to the UpP

D

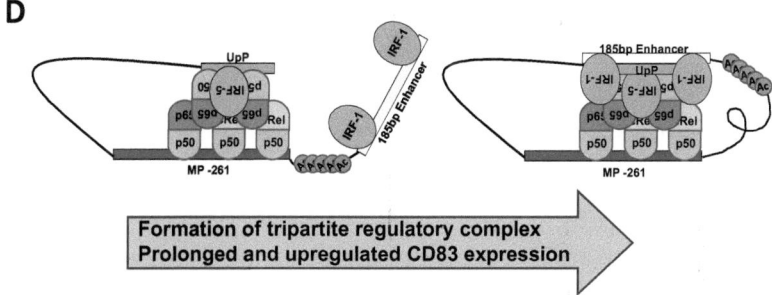

Formation of tripartite regulatory complex
Prolonged and upregulated CD83 expression

Figure 6.3. Proposed model of the CD83 promoter/enhancer complex regulating the cell type- and maturation status-specific CD83 expression in DCs. **(A)** Schematic depiction of the transcriptional regulation of the basal/transient CD83 expression in iDCs and non-DC cell types. The CD83 expression is solely driven by NFκB p50/p65 heterodimers inducing the CD83 minimal promoter (MP -261). IRF-2 constitutively binds to the enhancer, thereby blocking histone acetylation. **(B)** Upregulation of IRF-1, IRF-5 and cRel during DC maturation facilitates the upregulated and prolonged CD83 expression in mDCs. **(C)** NFκB p50/p65 heterodimers and IRF-5 cooperatively bind to the upstream promoter (UpP). IRF-2 is replaced by IRF-1 at the *185 bp enhancer*. IRF-1 recruits histone acetyltransferases (HAT) that acetylate (Ac) histones of CD83 intron 2 in a cell type- and maturation-specific manner. The transcription factor cRel replaces p65 at the MP -261. **(D)** With the now-bound IRF- and NFκB-factors and the eased DNA-bending, allowed by the histone acetylation, the ternary regulatory complex can form and induce an upregulated and prolonged CD83 expression.

Discussion

The basal or transient expression of CD83 in iDCs and non-DC cell types is mediated by NFκB p50/p65. Moreover, IRF-2 constitutively binds to the *185 bp enhancer* and blocks histone acetylation and therefore the functionality of the enhancer (Fig. 6.3. A). In DCs maturation stimuli like LPS lead to an activation of IRF-1, IRF-5 and cRel (Fig. 6.3. B). As a consequence, IRF-5 and NFκB p50/p65 can now bind cooperatively to the UpP. Moreover, cRel together with p50 might engage the NFκB-sites 4 and 5 (or replace p65), as cRel is strongly upregulated during DC maturation. Furthermore, IRF-1 replaces IRF-2 at the *185 bp enhancer*, recruiting HATs which acetylate histone 3 and lead to an open conformation of the chromatin (Fig. 6.3. C). Finally, a tripartite regulatory complex is formed, mediated by the interaction of the above mentioned transcription factors resulting in the looping of the UpP and the *185 bp enhancer* onto the MP -261 due to the acetylation. Thereby, the NFκB- and IRF-sites form *in trans* three IRF-NFκB modules that strongly upregulate and prolong the CD83 expression in mDCs.

Note that several aspects of the potential molecular mechanism governing the cell type- and maturation-specific CD83 expression have to be addressed in the future. Especially the molecular mechanisms that are involved in the CD83 expression in B and T cells have yet to be fully elucidated. A quantitative ChIP-assay for example could reveal the *in vivo* binding transcription factors in B and T cells, but also shed light on the binding of cRel and p65 to MP -261, as well as on the factors binding to NFκB-sites 1 and 2 in the UpP and the IRF-site 2 in the *185 bp enhancer*. Another interesting aspect that has to be addressed in the future is the physical interaction of NFκB – and IRF-factors (especially cRel and IRF-5) in DCs, which could be examined by coimmunoprecipitation experiments. Furthermore, the definite role of the SP1-sites in the MP –261 and the *185 bp enhancer* have yet to be elucidated by e.g. mutation experiments.

The full characterization of the CD83 promoter provides several benefits for upcoming experimental works in the fields of gene regulation, CD83 signaling and function as well as adenovirus-based DC vaccination therapies. The novel finding that IRF-NFκB modules can not only form *in cis*, but also *in trans*, extends the possibility of forming transcriptional modules in regulatory regions well beyond the well-known *cis*-modules. This might therefore enable genome-wide discoveries and the understanding of molecular mechanisms involved in

Discussion

transcriptional activation on a much broader basis than previously possible. Furthermore, the *in vivo* functions as well as the signaling of the membrane bound form of CD83 are still not fully elucidated. Revealing the regulation of this interesting molecule might provide useful insight into both topics.

Moreover, the characterization of the CD83 promoter not only allows a detailed look into the complex mechanisms of the immune response in general and the biology of DCs in particular, but it also exemplarily unfolds the cooperation of various promoter elements in a highly cell type- and status-specific immune cell related promoter.

Finally, for the first time a human mDC-specific promoter is described that might become an important tool for immunotherapeutic DC vaccination protocols. A fully characterized CD83 promoter including all regulatory elements contributing to the cell type and status specificity is a valuable tool to improve DC-mediated vaccination, as it provides the possibility to express multiple antigens (e.g. tumor, viral, bacterial) and immunomodulatory proteins, such as IL-12, specifically in maturing DCs. A mDC-specific promoter is the prerequisite to secure e.g. antigen presentation only in mDCs, but not in iDCs, thereby minimizing the risk of inducing tolerogenic mechanisms. In this context, the combination of the mDC-specific CD83 promoter and adenoviral gene therapy is a promising therapeutical approach, as the *in vitro* data indicate high functionality and specificity of the tripartite CD83 promoter/enhancer complex.

Reference List

1. Banchereau J, Steinman RM. Dendritic cells and the control of immunity. Nature 1998;392:245-252.

2. Steinman RM, Hawiger D, Nussenzweig MC. Tolerogenic dendritic cells. Annu.Rev.Immunol. 2003;21:685-711.

3. Steinman RM, Witmer MD. Lymphoid dendritic cells are potent stimulators of the primary mixed leukocyte reaction in mice. Proc.Natl.Acad.Sci.U.S.A 1978;75:5132-5136.

4. Nussenzweig MC, Steinman RM, Gutchinov B, Cohn ZA. Dendritic cells are accessory cells for the development of anti-trinitrophenyl cytotoxic T lymphocytes. J.Exp.Med. 1980;152:1070-1084.

5. Zhou LJ, Schwarting R, Smith HM, Tedder TF. A novel cell-surface molecule expressed by human interdigitating reticulum cells, Langerhans cells, and activated lymphocytes is a new member of the Ig superfamily. J.Immunol. 1992;149:735-742.

6. Lechmann M, Shuman N, Wakeham A, Mak TW. The CD83 reporter mouse elucidates the activity of the CD83 promoter in B, T, and dendritic cell populations in vivo. Proc.Natl.Acad.Sci.U.S.A 2008;105:11887-11892.

7. Schuurhuis DH, Fu N, Ossendorp F, Melief CJ. Ins and outs of dendritic cells. Int.Arch.Allergy Immunol. 2006;140:53-72.

8. Shortman K, Liu YJ. Mouse and human dendritic cell subtypes. Nat.Rev.Immunol. 2002;2:151-161.

9. Shortman K. Burnet oration: dendritic cells: multiple subtypes, multiple origins, multiple functions. Immunol.Cell Biol. 2000;78:161-165.

10. Wu L, D'Amico A, Hochrein H et al. Development of thymic and splenic dendritic cell populations from different hemopoietic precursors. Blood 2001;98:3376-3382.

11. Ardavin C. Origin, precursors and differentiation of mouse dendritic cells. Nat.Rev.Immunol. 2003;3:582-590.

12. Manz MG, Traver D, Miyamoto T, Weissman IL, Akashi K. Dendritic cell potentials of early lymphoid and myeloid progenitors. Blood 2001;97:3333-3341.

13. Traver D, Akashi K, Manz M et al. Development of CD8alpha-positive dendritic cells from a common myeloid progenitor. Science 2000;290:2152-2154.

14. Moser M, Murphy KM. Dendritic cell regulation of TH1-TH2 development. Nat.Immunol. 2000;1:199-205.

References

15. Pulendran B, Smith JL, Caspary G et al. Distinct dendritic cell subsets differentially regulate the class of immune response in vivo. Proc.Natl.Acad.Sci.U.S.A 1999;96:1036-1041.

16. Iyoda T, Shimoyama S, Liu K et al. The CD8+ dendritic cell subset selectively endocytoses dying cells in culture and in vivo. J.Exp.Med. 2002;195:1289-1302.

17. den Haan JM, Lehar SM, Bevan MJ. CD8(+) but not CD8(-) dendritic cells cross-prime cytotoxic T cells in vivo. J.Exp.Med. 2000;192:1685-1696.

18. Pooley JL, Heath WR, Shortman K. Cutting edge: intravenous soluble antigen is presented to CD4 T cells by CD8- dendritic cells, but cross-presented to CD8 T cells by CD8+ dendritic cells. J.Immunol. 2001;166:5327-5330.

19. Pulendran B, Banchereau J, Maraskovsky E, Maliszewski C. Modulating the immune response with dendritic cells and their growth factors. Trends Immunol. 2001;22:41-47.

20. De ST, Butz E, Smith J et al. CD8alpha(-) and CD8alpha(+) subclasses of dendritic cells undergo phenotypic and functional maturation in vitro and in vivo. J.Leukoc.Biol. 2001;69:951-958.

21. Reis e Sousa, Hieny S, Scharton-Kersten T et al. In vivo microbial stimulation induces rapid CD40 ligand-independent production of interleukin 12 by dendritic cells and their redistribution to T cell areas. J.Exp.Med. 1997;186:1819-1829.

22. Asselin-Paturel C, Boonstra A, Dalod M et al. Mouse type I IFN-producing cells are immature APCs with plasmacytoid morphology. Nat.Immunol. 2001;2:1144-1150.

23. Nakano H, Yanagita M, Gunn MD. CD11c(+)B220(+)Gr-1(+) cells in mouse lymph nodes and spleen display characteristics of plasmacytoid dendritic cells. J.Exp.Med. 2001;194:1171-1178.

24. Megjugorac NJ, Gallagher GE, Gallagher G. Modulation of human plasmacytoid DC function by IFN-lambda1 (IL-29). J.Leukoc.Biol. 2009;86:1359-1363.

25. Dallal RM, Lotze MT. The dendritic cell and human cancer vaccines. Curr.Opin.Immunol. 2000;12:583 588.

26. O'Doherty U, Peng M, Gezelter S et al. Human blood contains two subsets of dendritic cells, one immunologically mature and the other immature. Immunology 1994;82:487-493.

27. Ardavin C, Martinez del HG, Martin P et al. Origin and differentiation of dendritic cells. Trends Immunol. 2001;22:691-700.

28. Banchereau J, Pulendran B, Steinman R, Palucka K. Will the making of plasmacytoid dendritic cells in vitro help unravel their mysteries? J.Exp.Med. 2000;192:F39-F44.

29. Ito T, Liu YJ, Kadowaki N. Functional diversity and plasticity of human dendritic cell subsets. Int.J.Hematol. 2005;81:188-196.

30. Robinson SP, Patterson S, English N et al. Human peripheral blood contains two distinct lineages of dendritic cells. Eur.J.Immunol. 1999;29:2769-2778.

31. Caux C, Dezutter-Dambuyant C, Schmitt D, Banchereau J. GM-CSF and TNF-alpha cooperate in the generation of dendritic Langerhans cells. Nature 1992;360:258-261.

32. Nestle FO, Turka LA, Nickoloff BJ. Characterization of dermal dendritic cells in psoriasis. Autostimulation of T lymphocytes and induction of Th1 type cytokines. J.Clin.Invest 1994;94:202-209.

33. Caux C, Vanbervliet B, Massacrier C et al. CD34+ hematopoietic progenitors from human cord blood differentiate along two independent dendritic cell pathways in response to GM-CSF+TNF alpha. J.Exp.Med. 1996;184:695-706.

34. Valladeau J, Saeland S. Cutaneous dendritic cells. Semin.Immunol. 2005;17:273-283.

35. Banchereau J, Palucka AK. Dendritic cells as therapeutic vaccines against cancer. Nat.Rev.Immunol. 2005;5:296-306.

36. de Saint-Vis B, Fugier-Vivier I, Massacrier C et al. The cytokine profile expressed by human dendritic cells is dependent on cell subtype and mode of activation. J.Immunol. 1998;160:1666-1676.

37. Caux C, Massacrier C, Vanbervliet B et al. CD34+ hematopoietic progenitors from human cord blood differentiate along two independent dendritic cell pathways in response to granulocyte-macrophage colony-stimulating factor plus tumor necrosis factor alpha: II. Functional analysis. Blood 1997;90:1458-1470.

38. Dubois B, Massacrier C, Vanbervliet B et al. Critical role of IL-12 in dendritic cell-induced differentiation of naive B lymphocytes. J.Immunol. 1998;161:2223-2231.

39. Olweus J, BitMansour A, Warnke R et al. Dendritic cell ontogeny: a human dendritic cell lineage of myeloid origin. Proc.Natl.Acad.Sci.U.S.A 1997;94:12551-12556.

40. Strobl H, Scheinecker C, Riedl E et al. Identification of CD68+lin- peripheral blood cells with dendritic precursor characteristics. J.Immunol. 1998;161:740-748.

41. MacDonald KP, Munster DJ, Clark GJ et al. Characterization of human blood dendritic cell subsets. Blood 2002;100:4512-4520.

42. Ito T, Inaba M, Inaba K et al. A CD1a+/CD11c+ subset of human blood dendritic cells is a direct precursor of Langerhans cells. J.Immunol. 1999;163:1409-1419.

43. Siegal FP, Kadowaki N, Shodell M et al. The nature of the principal type 1 interferon-producing cells in human blood. Science 1999;284:1835-1837.

44. Grouard G, Rissoan MC, Filgueira L et al. The enigmatic plasmacytoid T cells develop into dendritic cells with interleukin (IL)-3 and CD40-ligand. J.Exp.Med. 1997;185:1101-1111.

45. Rissoan MC, Soumelis V, Kadowaki N et al. Reciprocal control of T helper cell and dendritic cell differentiation. Science 1999;283:1183-1186.

46. Ueno H, Klechevsky E, Morita R et al. Dendritic cell subsets in health and disease. Immunol.Rev. 2007;219:118-142.

47. Matsui T, Connolly JE, Michnevitz M et al. CD2 distinguishes two subsets of human plasmacytoid dendritic cells with distinct phenotype and functions. J.Immunol. 2009;182:6815-6823.

48. Bachem A, Guttler S, Hartung E et al. Superior antigen cross-presentation and XCR1 expression define human CD11c+CD141+ cells as homologues of mouse CD8+ dendritic cells. J.Exp.Med. 2010;207:1273-1281.

49. Crozat K, Guiton R, Contreras V et al. The XC chemokine receptor 1 is a conserved selective marker of mammalian cells homologous to mouse CD8alpha+ dendritic cells. J.Exp.Med. 2010;207:1283-1292.

50. Poulin LF, Salio M, Griessinger E et al. Characterization of human DNGR-1+ BDCA3+ leukocytes as putative equivalents of mouse CD8alpha+ dendritic cells. J.Exp.Med. 2010;207:1261-1271.

51. Stahl P, Schlesinger PH, Sigardson E, Rodman JS, Lee YC. Receptor-mediated pinocytosis of mannose glycoconjugates by macrophages: characterization and evidence for receptor recycling. Cell 1980;19:207-215.

52. Geijtenbeek TB, Torensma R, van Vliet SJ et al. Identification of DC-SIGN, a novel dendritic cell-specific ICAM-3 receptor that supports primary immune responses. Cell 2000;100:575-585.

53. Valladeau J, Ravel O, Dezutter-Dambuyant C et al. Langerin, a novel C-type lectin specific to Langerhans cells, is an endocytic receptor that induces the formation of Birbeck granules. Immunity. 2000;12:71-81.

54. Kato M, Neil TK, Clark GJ et al. cDNA cloning of human DEC-205, a putative antigen-uptake receptor on dendritic cells. Immunogenetics 1998;47:442-450.

55. Engering AJ, Cella M, Fluitsma DM et al. Mannose receptor mediated antigen uptake and presentation in human dendritic cells. Adv.Exp.Med.Biol. 1997;417:183-187.

56. Geijtenbeek TB, Krooshoop DJ, Bleijs DA et al. DC-SIGN-ICAM-2 interaction mediates dendritic cell trafficking. Nat.Immunol. 2000;1:353-357.

57. Maurer D, Fiebiger E, Reininger B et al. Fc epsilon receptor I on dendritic cells delivers IgE-bound multivalent antigens into a cathepsin S-dependent pathway of MHC class II presentation. J.Immunol. 1998;161:2731-2739.

58. Regnault A, Lankar D, Lacabanne V et al. Fcgamma receptor-mediated induction of dendritic cell maturation and major histocompatibility complex class I-restricted antigen presentation after immune complex internalization. J.Exp.Med. 1999;189:371-380.

59. Singh-Jasuja H, Hilf N, Arnold-Schild D, Schild H. The role of heat shock proteins and their receptors in the activation of the immune system. Biol.Chem. 2001;382:629-636.

60. Albert ML, Pearce SF, Francisco LM et al. Immature dendritic cells phagocytose apoptotic cells via alphavbeta5 and CD36, and cross-present antigens to cytotoxic T lymphocytes. J.Exp.Med. 1998;188:1359-1368.

61. Banchereau J, Steinman RM. Dendritic cells and the control of immunity. Nature 1998;392:245-252.

62. Charles A.Janeway, Paul Travers. Janeway Immunologie.: Current Biology Ltd./Garland Publishing Inc.; 1997.

63. Hunt DF, Michel H, Dickinson TA et al. Peptides presented to the immune system by the murine class II major histocompatibility complex molecule I-Ad. Science 1992;256:1817-1820.

64. Hunt DF, Henderson RA, Shabanowitz J et al. Characterization of peptides bound to the class I MHC molecule HLA-A2.1 by mass spectrometry. Science 1992;255:1261-1263.

65. Rudensky AY, Preston-Hurlburt P, al-Ramadi BK, Rothbard J, Janeway CA, Jr. Truncation variants of peptides isolated from MHC class II molecules suggest sequence motifs. Nature 1992;359:429-431.

66. Saudrais C, Spehner D, de la Salle H et al. Intracellular pathway for the generation of functional MHC class II peptide complexes in immature human dendritic cells. J.Immunol. 1998;160:2597-2607.

67. Cella M, Engering A, Pinet V, Pieters J, Lanzavecchia A. Inflammatory stimuli induce accumulation of MHC class II complexes on dendritic cells. Nature 1997;388:782-787.

68. Santambrogio L, Strominger JL. The ins and outs of MHC class II proteins in dendritic cells. Immunity. 2006;25:857-859.

69. Santambrogio L, Sato AK, Fischer FR, Dorf ME, Stern LJ. Abundant empty class II MHC molecules on the surface of immature dendritic cells. Proc.Natl.Acad.Sci.U.S.A 1999;96:15050-15055.

70. Santambrogio L, Sato AK, Carven GJ et al. Extracellular antigen processing and presentation by immature dendritic cells. Proc.Natl.Acad.Sci.U.S.A 1999;96:15056-15061.

71. Yewdell JW, Norbury CC, Bennink JR. Mechanisms of exogenous antigen presentation by MHC class I molecules in vitro and in vivo: implications for generating CD8+ T cell responses to infectious agents, tumors, transplants, and vaccines. Adv.Immunol. 1999;73:1-77.

72. Hofmann M, Nussbaum AK, Emmerich NP, Stoltze L, Schild H. Mechanisms of MHC class I-restricted antigen presentation. Expert.Opin.Ther.Targets. 2001;5:379-393.

73. Lehner PJ, Cresswell P. Recent developments in MHC-class-I-mediated antigen presentation. Curr.Opin.Immunol. 2004;16:82-89.

74. Rock KL, Shen L. Cross-presentation: underlying mechanisms and role in immune surveillance. Immunol.Rev. 2005;207:166-183.

75. Huang AY, Bruce AT, Pardoll DM, Levitsky HI. In vivo cross-priming of MHC class I-restricted antigens requires the TAP transporter. Immunity. 1996;4:349-355.

76. Blanchard N, Shastri N. Cross-presentation of peptides from intracellular pathogens by MHC class I molecules. Ann.N.Y.Acad.Sci. 2010;1183:237-250.

77. Belz GT, Carbone FR, Heath WR. Cross-presentation of antigens by dendritic cells. Crit Rev.Immunol. 2002;22:439-448.

78. Reis e Sousa. Toll-like receptors and dendritic cells: for whom the bug tolls. Semin.Immunol. 2004;16:27-34.

79. Takeda K, Akira S. Toll-like receptors in innate immunity. Int.Immunol. 2005;17:1-14.

80. Medzhitov R. Toll-like receptors and innate immunity. Nat.Rev.Immunol. 2001;1:135-145.

81. Akira S, Uematsu S, Takeuchi O. Pathogen recognition and innate immunity. Cell 2006;124:783-801.

82. Janeway CA, Jr., Medzhitov R. Innate immune recognition. Annu.Rev.Immunol. 2002;20:197-216.

83. Kaisho T, Akira S. Toll-like receptors as adjuvant receptors. Biochim.Biophys.Acta 2002;1589:1-13.

84. Kadowaki N, Ho S, Antonenko S et al. Subsets of human dendritic cell precursors express different toll-like receptors and respond to different microbial antigens. J.Exp.Med. 2001;194:863-869.

85. Hemmi H, Akira S. TLR signalling and the function of dendritic cells. Chem.Immunol.Allergy 2005;86:120-135.

86. Kawai T, Akira S. TLR signaling. Semin.Immunol. 2007;19:24-32.

87. O'Neill LA, Bowie AG. The family of five: TIR-domain-containing adaptors in Toll-like receptor signalling. Nat.Rev.Immunol. 2007;7:353-364.

88. Honda K, Taniguchi T. IRFs: master regulators of signalling by Toll-like receptors and cytosolic pattern-recognition receptors. Nat.Rev.Immunol. 2006;6:644-658.

89. Honda K, Taniguchi T. Toll-like receptor signaling and IRF transcription factors. IUBMB.Life 2006;58:290-295.

90. Macatonia SE, Hosken NA, Litton M et al. Dendritic cells produce IL-12 and direct the development of Th1 cells from naive CD4+ T cells. J.Immunol. 1995;154:5071-5079.

91. Heufler C, Koch F, Stanzl U et al. Interleukin-12 is produced by dendritic cells and mediates T helper 1 development as well as interferon-gamma production by T helper 1 cells. Eur.J.Immunol. 1996;26:659-668.

92. Schoenberger SP, Toes RE, van der Voort EI, Offringa R, Melief CJ. T-cell help for cytotoxic T lymphocytes is mediated by CD40-CD40L interactions. Nature 1998;393:480-483.

93. Grewal IS, Flavell RA. A central role of CD40 ligand in the regulation of CD4+ T-cell responses. Immunol.Today 1996;17:410-414.

94. Ridge JP, Di RF, Matzinger P. A conditioned dendritic cell can be a temporal bridge between a CD4+ T-helper and a T-killer cell. Nature 1998;393:474-478.

95. Diehl L, Den Boer AT, van der Voort EI et al. The role of CD40 in peripheral T cell tolerance and immunity. J.Mol.Med. 2000;78:363-371.

96. Schuurhuis DH, Ioan-Facsinay A, Nagelkerken B et al. Antigen-antibody immune complexes empower dendritic cells to efficiently prime specific CD8+ CTL responses in vivo. J.Immunol. 2002;168:2240-2246.

97. Enk AH. Dendritic cells in tolerance induction. Immunol.Lett. 2005;99:8-11.

98. Banchereau J, Briere F, Caux C et al. Immunobiology of dendritic cells. Annu.Rev.Immunol. 2000;18:767-811.

99. Caux C, Vanbervliet B, Massacrier C et al. Regulation of dendritic cell recruitment by chemokines. Transplantation 2002;73:S7-11.

100. Clark GJ, Angel N, Kato M et al. The role of dendritic cells in the innate immune system. Microbes.Infect. 2000;2:257-272.

101. Yanagihara S, Komura E, Nagafune J, Watarai H, Yamaguchi Y. EBI1/CCR7 is a new member of dendritic cell chemokine receptor that is up-regulated upon maturation. J.Immunol. 1998;161:3096-3102.

102. Yopp AC, Randolph GJ, Bromberg JS. Leukotrienes, sphingolipids, and leukocyte trafficking. J.Immunol. 2003;171:5-10.

103. Chan VW, Kothakota S, Rohan MC et al. Secondary lymphoid-tissue chemokine (SLC) is chemotactic for mature dendritic cells. Blood 1999;93:3610-3616.

104. O'Garra A, Murphy KM. From IL-10 to IL-12: how pathogens and their products stimulate APCs to induce T(H)1 development. Nat.Immunol. 2009;10:929-932.

105. O'Garra A, Murphy K. Role of cytokines in development of Th1 and Th2 cells. Chem.Immunol. 1996;63:1-13.

106. MacDonald AS, Maizels RM. Alarming dendritic cells for Th2 induction. J.Exp.Med. 2008;205:13-17.

107. Lutz MB, Kurts C. Induction of peripheral CD4+ T-cell tolerance and CD8+ T-cell cross-tolerance by dendritic cells. Eur.J.Immunol. 2009;39:2325-2330.

108. Matsushita S, Higashi T. Human Th17 cell clones and natural immune responses. Allergol.Int. 2008;57:135-140.

109. Fujita H, Nograles KE, Kikuchi T et al. Human Langerhans cells induce distinct IL-22-producing CD4+ T cells lacking IL-17 production. Proc.Natl.Acad.Sci.U.S.A 2009;106:21795-21800.

110. Behrens G, Li M, Smith CM et al. Helper T cells, dendritic cells and CTL Immunity. Immunol.Cell Biol. 2004;82:84-90.

111. Kidd P. Th1/Th2 balance: the hypothesis, its limitations, and implications for health and disease. Altern.Med.Rev. 2003;8:223-246.

112. Zhang S, Zhang H, Zhao J. The role of CD4 T cell help for CD8 CTL activation. Biochem.Biophys.Res.Commun. 2009;384:405-408.

113. Arnold B. Levels of peripheral T cell tolerance. Transpl.Immunol. 2002;10:109-114.

114. Wing K, Sakaguchi S. Regulatory T cells exert checks and balances on self tolerance and autoimmunity. Nat.Immunol. 2010;11:7-13.

115. Sakaguchi S, Wing K, Onishi Y, Prieto-Martin P, Yamaguchi T. Regulatory T cells: how do they suppress immune responses? Int.Immunol. 2009;21:1105-1111.

116. Kagami S, Rizzo HL, Lee JJ, Koguchi Y, Blauvelt A. Circulating Th17, Th22, and Th1 cells are increased in psoriasis. J.Invest Dermatol. 2010;130:1373-1383.

117. Bettelli E, Oukka M, Kuchroo VK. T(H)-17 cells in the circle of immunity and autoimmunity. Nat.Immunol. 2007;8:345-350.

118. Perona-Wright G, Jenkins SJ, O'Connor RA et al. A pivotal role for CD40-mediated IL-6 production by dendritic cells during IL-17 induction in vivo. J.Immunol. 2009;182:2808-2815.

119. Yssel H, Bensussan A. [Is there a novel subset of Th22 lymphocytes in the skin distinct from Th17 lymphocytes?]. Med.Sci.(Paris) 2010;26:12-14.

120. Zhou LJ, Tedder TF. Human blood dendritic cells selectively express CD83, a member of the immunoglobulin superfamily. J.Immunol. 1995;154:3821-3835.

121. Cao W, Lee SH, Lu J. CD83 is preformed inside monocytes, macrophages and dendritic cells, but it is only stably expressed on activated dendritic cells. Biochem.J. 2005;385:85-93.

122. Wolenski M, Cramer SO, Ehrlich S et al. Enhanced activation of CD83-positive T cells. Scand.J.Immunol. 2003;58:306-311.

123. Kozlow EJ, Wilson GL, Fox CH, Kehrl JH. Subtractive cDNA cloning of a novel member of the Ig gene superfamily expressed at high levels in activated B lymphocytes. Blood 1993;81:454-461.

124. Oehler L, Majdic O, Pickl WF et al. Neutrophil granulocyte-committed cells can be driven to acquire dendritic cell characteristics. J.Exp.Med. 1998;187:1019-1028.

125. Zhou LJ, Tedder TF. CD14+ blood monocytes can differentiate into functionally mature CD83+ dendritic cells. Proc.Natl.Acad.Sci.U.S.A 1996;93:2588-2592.

126. Iking-Konert C, Wagner C, Denefleh B et al. Up-regulation of the dendritic cell marker CD83 on polymorphonuclear neutrophils (PMN): divergent expression in acute bacterial infections and chronic inflammatory disease. Clin.Exp.Immunol. 2002;130:501-508.

127. Fujimoto Y, Tu L, Miller AS et al. CD83 expression influences CD4+ T cell development in the thymus. Cell 2002;108:755-767.

128. Sorg UR, Morse TM, Patton WN et al. Hodgkin's cells express CD83, a dendritic cell lineage associated antigen. Pathology 1997;29:294-299.

129. Dudziak D, Kieser A, Dirmeier U et al. Latent membrane protein 1 of Epstein-Barr virus induces CD83 by the NF-kappaB signaling pathway. J.Virol. 2003;77:8290-8298.

130. Lechmann M, Shuman N, Wakeham A, Mak TW. The CD83 reporter mouse elucidates the activity of the CD83 promoter in B, T, and dendritic cell populations in vivo. Proc.Natl.Acad.Sci.U.S.A 2008;105:11887-11892.

131. Wolenski M, Cramer SO, Ehrlich S et al. Expression of CD83 in the murine immune system. Med.Microbiol.Immunol. 2003;192:189-192.

132. Cramer SO, Trumpfheller C, Mehlhoop U et al. Activation-induced expression of murine CD83 on T cells and identification of a specific CD83 ligand on murine B cells. Int.Immunol. 2000;12:1347-1351.

133. Prechtel AT, Steinkasserer A. CD83: an update on functions and prospects of the maturation marker of dendritic cells. Arch.Dermatol.Res. 2007;299:59-69.

134. Rodriguez MS, Dargemont C, Stutz F. Nuclear export of RNA. Biol.Cell 2004;96:639-655.

135. Fries B, Heukeshoven J, Hauber I et al. Analysis of nucleocytoplasmic trafficking of the HuR ligand APRIL and its influence on CD83 expression. J.Biol.Chem. 2007;282:4504-4515.

136. Prechtel AT, Chemnitz J, Schirmer S et al. Expression of CD83 is regulated by HuR via a novel cis-active coding region RNA element. J.Biol.Chem. 2006;281:10912-10925.

137. Zinser E, Turza N, Steinkasserer A. CNI-1493 mediated suppression of dendritic cell activation in vitro and in vivo. Immunobiology 2004;209:89-97.

138. Hock BD, Kato M, McKenzie JL, Hart DN. A soluble form of CD83 is released from activated dendritic cells and B lymphocytes, and is detectable in normal human sera. Int.Immunol. 2001;13:959-967.

139. Hock BD, Haring LF, Steinkasserer A et al. The soluble form of CD83 is present at elevated levels in a number of hematological malignancies. Leuk.Res. 2004;28:237-241.

140. Hock BD, O'Donnell JL, Taylor K et al. Levels of the soluble forms of CD80, CD86, and CD83 are elevated in the synovial fluid of rheumatoid arthritis patients. Tissue Antigens 2006;67:57-60.

141. Dudziak D, Nimmerjahn F, Bornkamm GW, Laux G. Alternative splicing generates putative soluble CD83 proteins that inhibit T cell proliferation. J.Immunol. 2005;174:6672-6676.

142. Twist CJ, Beier DR, Disteche CM, Edelhoff S, Tedder TF. The mouse Cd83 gene: structure, domain organization, and chromosome localization. Immunogenetics 1998;48:383-393.

143. Hock BD, Kato M, McKenzie JL, Hart DN. A soluble form of CD83 is released from activated dendritic cells and B lymphocytes, and is detectable in normal human sera. Int.Immunol. 2001;13:959-967.

144. Berchtold S, Jones T, Muhl-Zurbes P et al. The human dendritic cell marker CD83 maps to chromosome 6p23. Ann.Hum.Genet. 1999;63:181-183.

145. Lechmann M, Kotzor N, Zinser E et al. CD83 is a dimer: Comparative analysis of monomeric and dimeric isoforms. Biochem.Biophys.Res.Commun. 2005;329:132-139.

146. Ohta Y, Landis E, Boulay T et al. Homologs of CD83 from elasmobranch and teleost fish. J.Immunol. 2004;173:4553-4560.

147. Lechmann M, Kotzor N, Zinser E et al. CD83 is a dimer: Comparative analysis of monomeric and dimeric isoforms. Biochem.Biophys.Res.Commun. 2005;329:132-139.

148. UniProt Consortium. Uniprot; Q01151 (CD83_Human). 2010. Ref Type: Online Source

149. Prechtel AT, Turza NM, Theodoridis AA, Steinkasserer A. CD83 knockdown in monocyte-derived dendritic cells by small interfering RNA leads to a diminished T cell stimulation. J.Immunol. 2007;178:5454-5464.

150. Aerts-Toegaert C, Heirman C, Tuyaerts S et al. CD83 expression on dendritic cells and T cells: correlation with effective immune responses. Eur.J.Immunol. 2007;37:686-695.

151. Kruse M, Rosorius O, Kratzer F et al. Inhibition of CD83 cell surface expression during dendritic cell maturation by interference with nuclear export of CD83 mRNA. J.Exp.Med. 2000;191:1581-1590.

152. Lechmann M, Krooshoop DJ, Dudziak D et al. The extracellular domain of CD83 inhibits dendritic cell-mediated T cell stimulation and binds to a ligand on dendritic cells. J.Exp.Med. 2001;194:1813-1821.

153. Kotzor N, Lechmann M, Zinser E, Steinkasserer A. The soluble form of CD83 dramatically changes the cytoskeleton of dendritic cells. Immunobiology 2004;209:129-140.

154. Scholler N, Hayden-Ledbetter M, Dahlin A et al. Cutting edge: CD83 regulates the development of cellular immunity. J.Immunol. 2002;168:2599-2602.

155. Zinser E, Lechmann M, Golka A, Lutz MB, Steinkasserer A. Prevention and treatment of experimental autoimmune encephalomyelitis by soluble CD83. J.Exp.Med. 2004;200:345-351.

156. Ge W, Arp J, Lian D et al. Immunosuppression involving soluble CD83 induces tolerogenic dendritic cells that prevent cardiac allograft rejection. Transplantation 2010;90:1145-1156.

157. Xu JF, Huang BJ, Yin H et al. A limited course of soluble CD83 delays acute cellular rejection of MHC-mismatched mouse skin allografts. Transpl.Int. 2007;20:266-276.

158. Lan Z, Lian D, Liu W et al. Prevention of chronic renal allograft rejection by soluble CD83. Transplantation 2010;90:1278-1285.

159. Lan Z, Ge W, Arp J et al. Induction of kidney allograft tolerance by soluble CD83 associated with prevalence of tolerogenic dendritic cells and indoleamine 2,3-dioxygenase. Transplantation 2010;90:1286-1293.

160. Hock BD, Fernyhough LJ, Gough SM et al. Release and clinical significance of soluble CD83 in chronic lymphocytic leukemia. Leuk.Res. 2009;33:1089-1095.

161. Scholler N, Hayden-Ledbetter M, Hellstrom KE, Hellstrom I, Ledbetter JA. CD83 is an I-type lectin adhesion receptor that binds monocytes and a subset of activated CD8+ T cells [corrected]. J.Immunol. 2001;166:3865-3872.

162. Hirano N, Butler MO, Xia Z et al. Engagement of CD83 ligand induces prolonged expansion of CD8+ T cells and preferential enrichment for antigen specificity. Blood 2006;107:1528-1536.

163. Senechal B, Boruchov AM, Reagan JL, Hart DN, Young JW. Infection of mature monocyte-derived dendritic cells with human cytomegalovirus inhibits stimulation of T-cell proliferation via the release of soluble CD83. Blood 2004;103:4207-4215.

164. Kruse M, Rosorius O, Kratzer F et al. Mature dendritic cells infected with herpes simplex virus type 1 exhibit inhibited T-cell stimulatory capacity. J.Virol. 2000;74:7127-7136.

165. Arrode G, Boccaccio C, Abastado JP, Davrinche C. Cross-presentation of human cytomegalovirus pp65 (UL83) to CD8+ T cells is regulated by virus-induced, soluble-mediator-dependent maturation of dendritic cells. J.Virol. 2002;76:142-150.

166. Raftery MJ, Schwab M, Eibert SM et al. Targeting the function of mature dendritic cells by human cytomegalovirus: a multilayered viral defense strategy. Immunity. 2001;15:997-1009.

167. Berchtold S, Muhl-Zurbes P, Maczek E et al. Cloning and characterization of the promoter region of the human CD83 gene. Immunobiology 2002;205:231-246.

168. McKinsey TA, Chu Z, Tedder TF, Ballard DW. Transcription factor NF-kappaB regulates inducible CD83 gene expression in activated T lymphocytes. Mol.Immunol. 2000;37:783-788.

169. Schierer S, Hesse A, Muller I et al. Modulation of viability and maturation of human monocyte-derived dendritic cells by oncolytic adenoviruses. Int.J.Cancer 2008;122:219-229.

170. Tomihara K, Kato K, Masuta Y et al. Gene transfer of CD40-ligand to dendritic cells stimulates interferon-gamma production to induce growth arrest and apoptosis of tumor cells. Gene Ther. 2008;15:203-213.

171. Wei J, Gao W, Wu J et al. Dendritic cells expressing a combined PADRE/MUC4-derived polyepitope DNA vaccine induce multiple cytotoxic T-cell responses. Cancer Biother Radiopharm. 2008;23:121-128.

172. McConnell MJ, Imperiale MJ. Biology of adenovirus and its use as a vector for gene therapy. Hum.Gene Ther. 2004;15:1022-1033.

173. Zhang Y, Bergelson JM. Adenovirus receptors. J.Virol. 2005;79:12125-12131.

174. Huang S, Endo RI, Nemerow GR. Upregulation of integrins alpha v beta 3 and alpha v beta 5 on human monocytes and T lymphocytes facilitates adenovirus-mediated gene delivery. J.Virol. 1995;69:2257-2263.

175. Wickham TJ, Mathias P, Cheresh DA, Nemerow GR. Integrins alpha v beta 3 and alpha v beta 5 promote adenovirus internalization but not virus attachment. Cell 1993;73:309-319.

176. Meier O, Boucke K, Hammer SV et al. Adenovirus triggers macropinocytosis and endosomal leakage together with its clathrin-mediated uptake. J.Cell Biol. 2002;158:1119-1131.

177. Stewart PL, Chiu CY, Huang S et al. Cryo-EM visualization of an exposed RGD epitope on adenovirus that escapes antibody neutralization. EMBO J. 1997;16:1189-1198.

178. Kelkar SA, Pfister KK, Crystal RG, Leopold PL. Cytoplasmic dynein mediates adenovirus binding to microtubules. J.Virol. 2004;78:10122-10132.

179. Trotman LC, Mosberger N, Fornerod M, Stidwill RP, Greber UF. Import of adenovirus DNA involves the nuclear pore complex receptor CAN/Nup214 and histone H1. Nat.Cell Biol. 2001;3:1092-1100.

180. Glasgow JN, Everts M, Curiel DT. Transductional targeting of adenovirus vectors for gene therapy. Cancer Gene Ther. 2006;13:830-844.

181. Xiang ZQ, Yang Y, Wilson JM, Ertl HC. A replication-defective human adenovirus recombinant serves as a highly efficacious vaccine carrier. Virology 1996;219:220-227.

182. Graham FL, Smiley J, Russell WC, Nairn R. Characteristics of a human cell line transformed by DNA from human adenovirus type 5. J.Gen.Virol. 1977;36:59-74.

183. Lochmuller H, Jani A, Huard J et al. Emergence of early region 1-containing replication-competent adenovirus in stocks of replication-defective adenovirus recombinants (delta E1 + delta E3) during multiple passages in 293 cells. Hum.Gene Ther. 1994;5:1485-1491.

184. Yang Y, Li Q, Ertl HC, Wilson JM. Cellular and humoral immune responses to viral antigens create barriers to lung-directed gene therapy with recombinant adenoviruses. J.Virol. 1995;69:2004-2015.

185. Yang Y, Ertl HC, Wilson JM. MHC class I-restricted cytotoxic T lymphocytes to viral antigens destroy hepatocytes in mice infected with E1-deleted recombinant adenoviruses. Immunity. 1994;1:433-442.

186. Amalfitano A, Hauser MA, Hu H et al. Production and characterization of improved adenovirus vectors with the E1, E2b, and E3 genes deleted. J.Virol. 1998;72:926-933.

187. Butterfield LH, Comin-Anduix B, Vujanovic L et al. Adenovirus MART-1-engineered autologous dendritic cell vaccine for metastatic melanoma. J.Immunother. 2008;31:294-309.

188. Hemminki A, Wang M, Desmond RA et al. Serum and ascites neutralizing antibodies in ovarian cancer patients treated with intraperitoneal adenoviral gene therapy. Hum.Gene Ther. 2002;13:1505-1514.

189. Okano M, Bell DW, Haber DA, Li E. DNA methyltransferases Dnmt3a and Dnmt3b are essential for de novo methylation and mammalian development. Cell 1999;99:247-257.

190. Li E, Beard C, Jaenisch R. Role for DNA methylation in genomic imprinting. Nature 1993;366:362-365.

191. Goto T, Monk M. Regulation of X-chromosome inactivation in development in mice and humans. Microbiol.Mol.Biol.Rev. 1998;62:362-378.

192. Walsh CP, Chaillet JR, Bestor TH. Transcription of IAP endogenous retroviruses is constrained by cytosine methylation. Nat.Genet. 1998;20:116-117.

193. Plass C, Soloway PD. DNA methylation, imprinting and cancer. Eur.J.Hum.Genet. 2002;10:6-16.

194. Herman JG, Baylin SB. Gene silencing in cancer in association with promoter hypermethylation. N.Engl.J.Med. 2003;349:2042-2054.

195. Ehrlich M, Gama-Sosa MA, Huang LH et al. Amount and distribution of 5-methylcytosine in human DNA from different types of tissues of cells. Nucleic Acids Res. 1982;10:2709-2721.

196. Vanyushin BF. Adenine Methylation in Eukaryotic DNA. Molecular Biology 2004;39:473-481.

197. Bolden AH, Nalin CM, Ward CA, Poonian MS, Weissbach A. Primary DNA sequence determines sites of maintenance and de novo methylation by mammalian DNA methyltransferases. Mol.Cell Biol. 1986;6:1135-1140.

198. Dodge JE, Ramsahoye BH, Wo ZG, Okano M, Li E. De novo methylation of MMLV provirus in embryonic stem cells: CpG versus non-CpG methylation. Gene 2002;289:41-48.

199. Lister R, Pelizzola M, Dowen RH et al. Human DNA methylomes at base resolution show widespread epigenomic differences. Nature 2009;462:315-322.

200. Haines TR, Rodenhiser DI, Ainsworth PJ. Allele-specific non-CpG methylation of the Nf1 gene during early mouse development. Dev.Biol. 2001;240:585-598.

201. Fuks F. DNA methylation and histone modifications: teaming up to silence genes. Curr.Opin.Genet.Dev. 2005;15:490-495.

202. Mohn F, Weber M, Rebhan M et al. Lineage-specific polycomb targets and de novo DNA methylation define restriction and potential of neuronal progenitors. Mol.Cell 2008;30:755-766.

203. Meissner A, Mikkelsen TS, Gu H et al. Genome-scale DNA methylation maps of pluripotent and differentiated cells. Nature 2008;454:766-770.

204. Metivier R, Gallais R, Tiffoche C et al. Cyclical DNA methylation of a transcriptionally active promoter. Nature 2008;452:45-50.

205. Tucker KL. Methylated cytosine and the brain: a new base for neuroscience. Neuron 2001;30:649-652.

206. Lander ES, Linton LM, Birren B et al. Initial sequencing and analysis of the human genome. Nature 2001;409:860-921.

207. Dodge JE, Okano M, Dick F et al. Inactivation of Dnmt3b in mouse embryonic fibroblasts results in DNA hypomethylation, chromosomal instability, and spontaneous immortalization. J.Biol.Chem. 2005;280:17986-17991.

208. Daura-Oller E, Cabre M, Montero MA, Paternain JL, Romeu A. Specific gene hypomethylation and cancer: new insights into coding region feature trends. Bioinformation. 2009;3:340-343.

209. Xu GL, Bestor TH, Bourc'his D et al. Chromosome instability and immunodeficiency syndrome caused by mutations in a DNA methyltransferase gene. Nature 1999;402:187-191.

210. Brero A, Easwaran HP, Nowak D et al. Methyl CpG-binding proteins induce large-scale chromatin reorganization during terminal differentiation. J.Cell Biol. 2005;169:733-743.

211. Okano M, Xie S, Li E. Dnmt2 is not required for de novo and maintenance methylation of viral DNA in embryonic stem cells. Nucleic Acids Res. 1998;26:2536-2540.

212. Reik W, Dean W. DNA methylation and mammalian epigenetics. Electrophoresis 2001;22:2838-2843.

213. Bestor TH. The DNA methyltransferases of mammals. Hum.Mol.Genet. 2000;9:2395-2402.

214. Jenuwein T, Allis CD. Translating the histone code. Science 2001;293:1074-1080.

215. Luger K, Richmond TJ. The histone tails of the nucleosome. Curr.Opin.Genet.Dev. 1998;8:140-146.

216. Kornberg RD, Lorch Y. Twenty-five years of the nucleosome, fundamental particle of the eukaryote chromosome. Cell 1999;98:285-294.

217. Jeon KW, Berezney R. *Structural and functional organization of the nuclear matrix*.; 1995.

218. Strahl BD, Allis CD. The language of covalent histone modifications. Nature 2000;403:41-45.

219. van Holde KE. Chromatin.; 1989.

220. Godde JS, Ura K. Cracking the enigmatic linker histone code. J.Biochem. 2008;143:287-293.

221. Xu L, Zhao Z, Dong A et al. Di- and tri- but not monomethylation on histone H3 lysine 36 marks active transcription of genes involved in flowering time regulation and other processes in Arabidopsis thaliana. Mol.Cell Biol. 2008;28:1348-1360.

222. Kubicek S, Jenuwein T. A crack in histone lysine methylation. Cell 2004;119:903-906.

223. Nimura K, Ura K, Kaneda Y. Histone methyltransferases: regulation of transcription and contribution to human disease. J.Mol.Med. 2010

224. Tsukada Y, Fang J, Erdjument-Bromage H et al. Histone demethylation by a family of JmjC domain-containing proteins. Nature 2006;439:811-816.

225. Shi Y, Lan F, Matson C et al. Histone demethylation mediated by the nuclear amine oxidase homolog LSD1. Cell 2004;119:941-953.

226. Tian X, Fang J. Current perspectives on histone demethylases. Acta Biochim.Biophys.Sin.(Shanghai) 2007;39:81-88.

227. Turner BM. Histone acetylation and an epigenetic code. Bioessays 2000;22:836-845.

228. Takahashi H, McCaffery JM, Irizarry RA, Boeke JD. Nucleocytosolic acetyl-coenzyme a synthetase is required for histone acetylation and global transcription. Mol.Cell 2006;23:207-217.

229. Li Q, Barkess G, Qian H. Chromatin looping and the probability of transcription. Trends Genet. 2006;22:197-202.

230. Petrascheck M, Escher D, Mahmoudi T et al. DNA looping induced by a transcriptional enhancer in vivo. Nucleic Acids Res. 2005;33:3743-3750.

231. Yang XJ, Seto E. HATs and HDACs: from structure, function and regulation to novel strategies for therapy and prevention. Oncogene 2007;26:5310-5318.

232. Hassig CA, Schreiber SL. Nuclear histone acetylases and deacetylases and transcriptional regulation: HATs off to HDACs. Curr.Opin.Chem.Biol. 1997;1:300-308.

233. Smale ST, Kadonaga JT. The RNA polymerase II core promoter. Annu.Rev.Biochem. 2003;72:449-479.

234. Rolf Knippers. Molekulare Genetik.: Georg Thieme Verlag; 2001.

235. Maston GA, Evans SK, Green MR. Transcriptional regulatory elements in the human genome. Annu.Rev.Genomics Hum.Genet. 2006;7:29-59.

236. Kim S, Na JG, Hampsey M, Reinberg D. The Dr1/DRAP1 heterodimer is a global repressor of transcription in vivo. Proc.Natl.Acad.Sci.U.S.A 1997;94:820-825.

References

237. Orphanides G, Lagrange T, Reinberg D. The general transcription factors of RNA polymerase II. Genes Dev. 1996;10:2657-2683.

238. Vilar JM, Saiz L. DNA looping in gene regulation: from the assembly of macromolecular complexes to the control of transcriptional noise. Curr.Opin.Genet.Dev. 2005;15:136-144.

239. Privalsky ML. The role of corepressors in transcriptional regulation by nuclear hormone receptors. Annu.Rev.Physiol 2004;66:315-360.

240. Gershenzon NI, Ioshikhes IP. Synergy of human Pol II core promoter elements revealed by statistical sequence analysis. Bioinformatics. 2005;21:1295-1300.

241. Lee TI, Young RA. Transcription of eukaryotic protein-coding genes. Annu.Rev.Genet. 2000;34:77-137.

242. Conaway JW, Florens L, Sato S et al. The mammalian Mediator complex. FEBS Lett. 2005;579:904-908.

243. Lee DH, Gershenzon N, Gupta M et al. Functional characterization of core promoter elements: the downstream core element is recognized by TAF1. Mol.Cell Biol. 2005;25:9674-9686.

244. Lim CY, Santoso B, Boulay T et al. The MTE, a new core promoter element for transcription by RNA polymerase II. Genes Dev. 2004;18:1606-1617.

245. Hahn S. Structure and mechanism of the RNA polymerase II transcription machinery. Nat.Struct.Mol.Biol. 2004;11:394-403.

246. Chen Z, Manley JL. Core promoter elements and TAFs contribute to the diversity of transcriptional activation in vertebrates. Mol.Cell Biol. 2003;23:7350-7362.

247. Claessens F, Gewirth DT. DNA recognition by nuclear receptors. Essays Biochem. 2004;40:59-72.

248. Carey M., Smale ST. Transcriptional Regulation in Eukaryotes: Concepts, Strategies, and Techniques .; 2000.

249. Massari ME, Murre C. Helix-loop-helix proteins: regulators of transcription in eucaryotic organisms. Mol.Cell Biol. 2000;20:429-440.

250. Pabo CO, Sauer RT. Transcription factors: structural families and principles of DNA recognition. Annu.Rev.Biochem. 1992;61:1053-1095.

251. Lonard DM, O'Malley BW. Expanding functional diversity of the coactivators. Trends Biochem.Sci. 2005;30:126-132.

252. Spiegelman BM, Heinrich R. Biological control through regulated transcriptional coactivators. Cell 2004;119:157-167.

253. Lemon B, Tjian R. Orchestrated response: a symphony of transcription factors for gene control. Genes Dev. 2000;14:2551-2569.

254. De la Serna IL, Ohkawa Y, Berkes CA et al. MyoD targets chromatin remodeling complexes to the myogenin locus prior to forming a stable DNA-bound complex. Mol.Cell Biol. 2005;25:3997-4009.

255. Ioshikhes IP, Zhang MQ. Large-scale human promoter mapping using CpG islands. Nat.Genet. 2000;26:61-63.

256. Blackwood EM, Kadonaga JT. Going the distance: a current view of enhancer action. Science 1998;281:60-63.

257. Lettice LA, Heaney SJ, Purdie LA et al. A long-range Shh enhancer regulates expression in the developing limb and fin and is associated with preaxial polydactyly. Hum.Mol.Genet. 2003;12:1725-1735.

258. Peyton DK, Ramesh T, Spear BT. Position-dependent activity of alpha - fetoprotein enhancer element III in the adult liver is due to negative regulation. Proc.Natl.Acad.Sci.U.S.A 2000;97:10890-10894.

259. Remenyi A, Scholer HR, Wilmanns M. Combinatorial control of gene expression. Nat.Struct.Mol.Biol. 2004;11:812-815.

260. Thanos D, Maniatis T. Virus induction of human IFN beta gene expression requires the assembly of an enhanceosome. Cell 1995;83:1091-1100.

261. Szutorisz H, Dillon N, Tora L. The role of enhancers as centres for general transcription factor recruitment. Trends Biochem.Sci. 2005;30:593-599.

262. Arnosti DN, Kulkarni MM. Transcriptional enhancers: Intelligent enhanceosomes or flexible billboards? J.Cell Biochem. 2005;94:890-898.

263. Arnosti DN. Analysis and function of transcriptional regulatory elements: insights from Drosophila. Annu.Rev.Entomol. 2003;48:579-602.

264. Kulkarni MM, Arnosti DN. Information display by transcriptional enhancers. Development 2003;130:6569-6575.

265. Ogbourne S, Antalis TM. Transcriptional control and the role of silencers in transcriptional regulation in eukaryotes. Biochem.J. 1998;331 (Pt 1):1-14.

266. Harris MB, Mostecki J, Rothman PB. Repression of an interleukin-4-responsive promoter requires cooperative BCL-6 function. J.Biol.Chem. 2005;280:13114-13121.

267. Sertil O, Kapoor R, Cohen BD, Abramova N, Lowry CV. Synergistic repression of anaerobic genes by Mot3 and Rox1 in Saccharomyces cerevisiae. Nucleic Acids Res. 2003;31:5831-5837.

268. Schmitt S, Prestel M, Paro R. Intergenic transcription through a polycomb group response element counteracts silencing. Genes Dev. 2005;19:697-708.

269. Recillas-Targa F, Pikaart MJ, Burgess-Beusse B et al. Position-effect protection and enhancer blocking by the chicken beta-globin insulator are separable activities. Proc.Natl.Acad.Sci.U.S.A 2002;99:6883-6888.

270. Felsenfeld G, Burgess-Beusse B, Farrell C et al. Chromatin boundaries and chromatin domains. Cold Spring Harb.Symp.Quant.Biol. 2004;69:245-250.

271. Li Q, Zhang M, Duan Z, Stamatoyannopoulos G. Structural analysis and mapping of DNase I hypersensitivity of HS5 of the beta-globin locus control region. Genomics 1999;61:183-193.

272. Capelson M, Corces VG. Boundary elements and nuclear organization. Biol.Cell 2004;96:617-629.

273. Defossez PA, Kelly KF, Filion GJ et al. The human enhancer blocker CTC-binding factor interacts with the transcription factor Kaiso. J.Biol.Chem. 2005;280:43017-43023.

274. Venter JC, Adams MD, Myers EW et al. The sequence of the human genome. Science 2001;291:1304-1351.

275. Elefant F, Cooke NE, Liebhaber SA. Targeted recruitment of histone acetyltransferase activity to a locus control region. J.Biol.Chem. 2000;275:13827-13834.

276. Ge Z, Li W, Wang N et al. Chromatin remodeling: recruitment of histone demethylase RBP2 by Mad1 for transcriptional repression of a Myc target gene, telomerase reverse transcriptase. FASEB J. 2010;24:579-586.

277. Perissi V, Aggarwal A, Glass CK, Rose DW, Rosenfeld MG. A corepressor/coactivator exchange complex required for transcriptional activation by nuclear receptors and other regulated transcription factors. Cell 2004;116:511-526.

278. Wierstra I. Sp1: emerging roles--beyond constitutive activation of TATA-less housekeeping genes. Biochem.Biophys.Res.Commun. 2008;372:1-13.

279. Li L, He S, Sun JM, Davie JR. Gene regulation by Sp1 and Sp3. Biochem.Cell Biol. 2004;82:460-471.

280. Hayden MS, West AP, Ghosh S. NF-kappaB and the immune response. Oncogene 2006;25:6758-6780.

281. Savitsky D, Tamura T, Yanai H, Taniguchi T. Regulation of immunity and oncogenesis by the IRF transcription factor family. Cancer Immunol.Immunother. 2010;59:489-510.

282. Dynan WS, Tjian R. Isolation of transcription factors that discriminate between different promoters recognized by RNA polymerase II. Cell 1983;32:669-680.

283. Kaczynski J, Cook T, Urrutia R. Sp1- and Kruppel-like transcription factors. Genome Biol. 2003;4:206.

284. Santiago FS, Ishii H, Shafi S et al. Yin Yang-1 inhibits vascular smooth muscle cell growth and intimal thickening by repressing p21WAF1/Cip1 transcription and p21WAF1/Cip1-Cdk4-cyclin D1 assembly. Circ.Res. 2007;101:146-155.

285. Opitz OG, Rustgi AK. Interaction between Sp1 and cell cycle regulatory proteins is important in transactivation of a differentiation-related gene. Cancer Res. 2000;60:2825-2830.

286. Mazure NM, Brahimi-Horn MC, Pouyssegur J. Protein kinases and the hypoxia-inducible factor-1, two switches in angiogenesis. Curr.Pharm.Des 2003;9:531-541.

287. Jones KA, Kadonaga JT, Luciw PA, Tjian R. Activation of the AIDS retrovirus promoter by the cellular transcription factor, Sp1. Science 1986;232:755-759.

288. Briggs MR, Kadonaga JT, Bell SP, Tjian R. Purification and biochemical characterization of the promoter-specific transcription factor, Sp1. Science 1986;234:47-52.

289. Kadonaga JT, Carner KR, Masiarz FR, Tjian R. Isolation of cDNA encoding transcription factor Sp1 and functional analysis of the DNA binding domain. Cell 1987;51:1079-1090.

290. Kadonaga JT, Tjian R. Affinity purification of sequence-specific DNA binding proteins. Proc.Natl.Acad.Sci.U.S.A 1986;83:5889-5893.

291. Olofsson BA, Kelly CM, Kim J, Hornsby SM, Azizkhan-Clifford J. Phosphorylation of Sp1 in response to DNA damage by ataxia telangiectasia-mutated kinase. Mol.Cancer Res. 2007;5:1319-1330.

292. Suzuki T, Kimura A, Nagai R, Horikoshi M. Regulation of interaction of the acetyltransferase region of p300 and the DNA-binding domain of Sp1 on and through DNA binding. Genes Cells 2000;5:29-41.

293. Zhao S, Venkatasubbarao K, Li S, Freeman JW. Requirement of a specific Sp1 site for histone deacetylase-mediated repression of transforming growth factor beta Type II receptor expression in human pancreatic cancer cells. Cancer Res. 2003;63:2624-2630.

294. Tone M, Tone Y, Babik JM, Lin CY, Waldmann H. The role of Sp1 and NF-kappa B in regulating CD40 gene expression. J.Biol.Chem. 2002;277:8890-8897.

295. Sen R, Baltimore D. Multiple nuclear factors interact with the immunoglobulin enhancer sequences. Cell 1986;46:705 716.

296. Breitman ML, Lai MM, Vogt PK. The genomic RNA of avian reticuloendotheliosis virus REV. Virology 1980;100:450-461.

297. Galas DJ, Schmitz A. DNAse footprinting: a simple method for the detection of protein-DNA binding specificity. Nucleic Acids Res. 1978;5:3157-3170.

298. Gilmore TD, Temin HM. Different localization of the product of the v-rel oncogene in chicken fibroblasts and spleen cells correlates with transformation by REV-T. Cell 1986;44:791-800.

References

299. Sen R, Baltimore D. Inducibility of kappa immunoglobulin enhancer-binding protein Nf-kappa B by a posttranslational mechanism. Cell 1986;47:921-928.

300. Baeuerle PA, Baltimore D. Activation of DNA-binding activity in an apparently cytoplasmic precursor of the NF-kappa B transcription factor. Cell 1988;53:211-217.

301. Steward R, Zusman SB, Huang LH, Schedl P. The dorsal protein is distributed in a gradient in early Drosophila embryos. Cell 1988;55:487-495.

302. Stephens RM, Rice NR, Hiebsch RR, Bose HR, Jr., Gilden RV. Nucleotide sequence of v-rel: the oncogene of reticuloendotheliosis virus. Proc.Natl.Acad.Sci.U.S.A 1983;80:6229-6233.

303. Wilhelmsen KC, Eggleton K, Temin HM. Nucleic acid sequences of the oncogene v-rel in reticuloendotheliosis virus strain T and its cellular homolog, the proto-oncogene c-rel. J.Virol. 1984;52:172-182.

304. Steward R. Dorsal, an embryonic polarity gene in Drosophila, is homologous to the vertebrate proto-oncogene, c-rel. Science 1987;238:692-694.

305. Ghosh S, Gifford AM, Riviere LR et al. Cloning of the p50 DNA binding subunit of NF-kappa B: homology to rel and dorsal. Cell 1990;62:1019-1029.

306. Kieran M, Blank V, Logeat F et al. The DNA binding subunit of NF-kappa B is identical to factor KBF1 and homologous to the rel oncogene product. Cell 1990;62:1007-1018.

307. Mebius RE. Organogenesis of lymphoid tissues. Nat.Rev.Immunol. 2003;3:292-303.

308. Bonizzi G, Karin M. The two NF-kappaB activation pathways and their role in innate and adaptive immunity. Trends Immunol. 2004;25:280-288.

309. Hayden MS, Ghosh S. Signaling to NF-kappaB. Genes Dev. 2004;18:2195-2224.

310. Lin X, Wang D. The roles of CARMA1, Bcl10, and MALT1 in antigen receptor signaling. Semin.Immunol. 2004;16:429-435.

311. Siebenlist U, Brown K, Claudio E. Control of lymphocyte development by nuclear factor-kappaB. Nat.Rev.Immunol. 2005;5:435-445.

312. Gilmore TD. Introduction to NF-kappaB: players, pathways, perspectives. Oncogene 2006;25:6680-6684.

313. Parul Tripathi AA. NF-kB transcription factor: a key player in generation of immune response. Current Science 2006;90:519-531.

314. Murphy TL, Cleveland MG, Kulesza P, Magram J, Murphy KM. Regulation of interleukin 12 p40 expression through an NF-kappa B half-site. Mol.Cell Biol. 1995;15:5258-5267.

315. Tas SW, Vervoordeldonk MJ, Hajji N et al. Noncanonical NF-kappaB signaling in dendritic cells is required for indoleamine 2,3-dioxygenase (IDO) induction and immune regulation. Blood 2007;110:1540-1549.

316. Gilmore TD, Kalaitzidis D, Liang MC, Starczynowski DT. The c-Rel transcription factor and B-cell proliferation: a deal with the devil. Oncogene 2004;23:2275-2286.

317. Ghosh S, May MJ, Kopp EB. NF-kappa B and Rel proteins: evolutionarily conserved mediators of immune responses. Annu.Rev.Immunol. 1998;16:225-260.

318. Fan CM, Maniatis T. Generation of p50 subunit of NF-kappa B by processing of p105 through an ATP-dependent pathway. Nature 1991;354:395-398.

319. Mercurio F, Didonato J, Rosette C, Karin M. Molecular cloning and characterization of a novel Rel/NF-kappa B family member displaying structural and functional homology to NF-kappa B p50/p105. DNA Cell Biol. 1992;11:523-537.

320. Jacobs MD, Harrison SC. Structure of an IkappaBalpha/NF-kappaB complex. Cell 1998;95:749-758.

321. Verma IM, Stevenson JK, Schwarz EM, Van AD, Miyamoto S. Rel/NF-kappa B/I kappa B family: intimate tales of association and dissociation. Genes Dev. 1995;9:2723-2735.

322. Gilmore TD, Morin PJ. The I kappa B proteins: members of a multifunctional family. Trends Genet. 1993;9:427-433.

323. May MJ, Ghosh S. Rel/NF-kappa B and I kappa B proteins: an overview. Semin.Cancer Biol. 1997;8:63-73.

324. Prajapati S, Gaynor RB. Regulation of Ikappa B kinase (IKK)gamma /NEMO function by IKKbeta -mediated phosphorylation. J.Biol.Chem. 2002;277:24331-24339.

325. Yamamoto Y, Kim DW, Kwak YT et al. IKKgamma /NEMO facilitates the recruitment of the IkappaB proteins into the IkappaB kinase complex. J.Biol.Chem. 2001;276:36327-36336.

326. Hacker H, Karin M. Regulation and function of IKK and IKK-related kinases. Sci.STKE. 2006;2006:re13.

327. May MJ, Ghosh S. IkappaB kinases: kinsmen with different crafts. Science 1999;284:271-273.

328. Paria BC, Malik AB, Kwiatek AM et al. Tumor necrosis factor-alpha induces nuclear factor-kappaB-dependent TRPC1 expression in endothelial cells. J.Biol.Chem. 2003;278:37195-37203.

329. Fitzgerald KA, Rowe DC, Barnes BJ et al. LPS-TLR4 signaling to IRF-3/7 and NF-kappaB involves the toll adapters TRAM and TRIF. J.Exp.Med. 2003;198:1043-1055.

330. May MJ, Ghosh S. Signal transduction through NF-kappa B. Immunol.Today 1998;19:80-88.

331. May MJ, Ghosh S. Rel/NF-kappa B and I kappa B proteins: an overview. Semin.Cancer Biol. 1997;8:63-73.

332. Ghosh S, Karin M. Missing pieces in the NF-kappaB puzzle. Cell 2002;109 Suppl:S81-S96.

333. Granelli-Piperno A, Pope M, Inaba K, Steinman RM. Coexpression of NF-kappa B/Rel and Sp1 transcription factors in human immunodeficiency virus 1-induced, dendritic cell-T-cell syncytia. Proc.Natl.Acad.Sci.U.S.A 1995;92:10944-10948.

334. Karin M, Ben-Neriah Y. Phosphorylation meets ubiquitination: the control of NF-[kappa]B activity. Annu.Rev.Immunol. 2000;18:621-663.

335. Burkly L, Hession C, Ogata L et al. Expression of relB is required for the development of thymic medulla and dendritic cells. Nature 1995;373:531-536.

336. Ouaaz F, Arron J, Zheng Y, Choi Y, Beg AA. Dendritic cell development and survival require distinct NF-kappaB subunits. Immunity. 2002;16:257-270.

337. Burkly L, Hession C, Ogata L et al. Expression of relB is required for the development of thymic medulla and dendritic cells. Nature 1995;373:531-536.

338. Martin E, O'Sullivan B, Low P, Thomas R. Antigen-specific suppression of a primed immune response by dendritic cells mediated by regulatory T cells secreting interleukin-10. Immunity. 2003;18:155-167.

339. Li M, Zhang X, Zheng X et al. Immune modulation and tolerance induction by RelB-silenced dendritic cells through RNA interference. J.Immunol. 2007;178:5480-5487.

340. Platzer B, Jorgl A, Taschner S, Hocher B, Strobl H. RelB regulates human dendritic cell subset development by promoting monocyte intermediates. Blood 2004;104:3655-3663.

341. Yang H, Zhang Y, Wu M et al. Suppression of ongoing experimental autoimmune myasthenia gravis by transfer of RelB-silenced bone marrow dentritic cells is associated with a change from a T helper Th17/Th1 to a Th2 and FoxP3+ regulatory T-cell profile. Inflamm.Res. 2010;59:197-205.

342. van de Laar L, van den Bosch A, van der Kooij SW et al. A nonredundant role for canonical NF-kappaB in human myeloid dendritic cell development and function. J.Immunol. 2010;185:7252-7261.

343. Neumann M, Fries H, Scheicher C et al. Differential expression of Rel/NF-kappaB and octamer factors is a hallmark of the generation and maturation of dendritic cells. Blood 2000;95:277-285.

344. Wang LC, Lin YL, Liang YC et al. The effect of caffeic acid phenethyl ester on the functions of human monocyte-derived dendritic cells. BMC.Immunol. 2009;10:39.

345. Zhou LF, Zhang MS, Yin KS et al. Effects of adenoviral gene transfer of mutated IkappaBalpha, a novel inhibitor of NF-kappaB, on human monocyte-derived dendritic cells. Acta Pharmacol.Sin. 2006;27:609-616.

346. Hernandez A, Burger M, Blomberg BB et al. Inhibition of NF-kappa B during human dendritic cell differentiation generates anergy and regulatory T-cell activity for one but not two human leukocyte antigen DR mismatches. Hum.Immunol. 2007;68:715-729.

347. Moynagh PN. TLR signalling and activation of IRFs: revisiting old friends from the NF-kappaB pathway. Trends Immunol. 2005;26:469-476.

348. Martin E, O'Sullivan B, Low P, Thomas R. Antigen-specific suppression of a primed immune response by dendritic cells mediated by regulatory T cells secreting interleukin-10. Immunity. 2003;18:155-167.

349. Harada H, Fujita T, Miyamoto M et al. Structurally similar but functionally distinct factors, IRF-1 and IRF-2, bind to the same regulatory elements of IFN and IFN-inducible genes. Cell 1989;58:729-739.

350. Taniguchi T, Ogasawara K, Takaoka A, Tanaka N. IRF family of transcription factors as regulators of host defense. Annu.Rev.Immunol. 2001;19:623-655.

351. Tamura T, Yanai H, Savitsky D, Taniguchi T. The IRF family transcription factors in immunity and oncogenesis. Annu.Rev.Immunol. 2008;26:535-584.

352. Alter-Koltunoff M, Ehrlich S, Dror N et al. Nramp1-mediated innate resistance to intraphagosomal pathogens is regulated by IRF-8, PU.1, and Miz-1. J.Biol.Chem. 2003;278:44025-44032.

353. Nehyba J, Hrdlickova R, Burnside J, Bose HR, Jr. A novel interferon regulatory factor (IRF), IRF-10, has a unique role in immune defense and is induced by the v-Rel oncoprotein. Mol.Cell Biol. 2002;22:3942-3957.

354. Huang B, Qi ZT, Xu Z, Nie P. Global characterization of interferon regulatory factor (IRF) genes in vertebrates: glimpse of the diversification in evolution. BMC.Immunol. 2010;11:22.

355. Battistini A. Interferon regulatory factors in hematopoietic cell differentiation and immune regulation. J.Interferon Cytokine Res. 2009;29:765-780.

356. Kato H, Sato S, Yoneyama M et al. Cell type-specific involvement of RIG-I in antiviral response. Immunity. 2005;23:19-28.

357. Kawai T, Takahashi K, Sato S et al. IPS-1, an adaptor triggering RIG-I- and Mda5-mediated type I interferon induction. Nat.Immunol. 2005;6:981-988.

358. Yoneyama M, Kikuchi M, Matsumoto K et al. Shared and unique functions of the DExD/H-box helicases RIG-I, MDA5, and LGP2 in antiviral innate immunity. J.Immunol. 2005;175:2851-2858.

359. Li K, Chen Z, Kato N, Gale M, Jr., Lemon SM. Distinct poly(I-C) and virus-activated signaling pathways leading to interferon-beta production in hepatocytes. J.Biol.Chem. 2005;280:16739-16747.

360. Gale M, Jr., Foy EM. Evasion of intracellular host defence by hepatitis C virus. Nature 2005;436:939-945.

361. Perry AK, Chen G, Zheng D, Tang H, Cheng G. The host type I interferon response to viral and bacterial infections. Cell Res. 2005;15:407-422.

362. Schroder M, Bowie AG. TLR3 in antiviral immunity: key player or bystander? Trends Immunol. 2005;26:462-468.

363. Kariko K, Buckstein M, Ni H, Weissman D. Suppression of RNA recognition by Toll-like receptors: the impact of nucleoside modification and the evolutionary origin of RNA. Immunity. 2005;23:165-175.

364. Honda K, Yanai H, Negishi H et al. IRF-7 is the master regulator of type-I interferon-dependent immune responses. Nature 2005;434:772-777.

365. Rifkin IR, Leadbetter EA, Busconi L, Viglianti G, Marshak-Rothstein A. Toll-like receptors, endogenous ligands, and systemic autoimmune disease. Immunol.Rev. 2005;204:27-42.

366. Haller O, Weber F. The interferon response circuit in antiviral host defense. Verh.K.Acad.Geneeskd.Belg. 2009;71:73-86.

367. Kroger A, Koster M, Schroeder K, Hauser H, Mueller PP. Activities of IRF-1. J.Interferon Cytokine Res. 2002;22:5-14.

368. Ohmori Y, Hamilton TA. IL-4-induced STAT6 suppresses IFN-gamma-stimulated STAT1-dependent transcription in mouse macrophages. J.Immunol. 1997;159:5474-5482.

369. Ohmori Y, Schreiber RD, Hamilton TA. Synergy between interferon-gamma and tumor necrosis factor-alpha in transcriptional activation is mediated by cooperation between signal transducer and activator of transcription 1 and nuclear factor kappaB. J.Biol.Chem. 1997;272:14899-14907.

370. Harada H, Kitagawa M, Tanaka N et al. Anti-oncogenic and oncogenic potentials of interferon regulatory factors-1 and -2. Science 1993;259:971-974.

371. Stevens AM, Yu-Lee LY. The transcription factor interferon regulatory factor-1 is expressed during both early G1 and the G1/S transition in the prolactin-induced lymphocyte cell cycle. Mol.Endocrinol. 1992;6:2236-2243.

372. Coccia EM, Stellacci E, Marziali G, Weiss G, Battistini A. IFN-gamma and IL-4 differently regulate inducible NO synthase gene expression through IRF-1 modulation. Int.Immunol. 2000;12:977-985.

373. Schaper F, Kirchhoff S, Posern G et al. Functional domains of interferon regulatory factor I (IRF-1). Biochem.J. 1998;335 (Pt 1):147-157.

374. Zhao J, Kong HJ, Li H et al. IRF-8/interferon (IFN) consensus sequence-binding protein is involved in Toll-like receptor (TLR) signaling and contributes to the cross-talk between TLR and IFN-gamma signaling pathways. J.Biol.Chem. 2006;281:10073-10080.

375. Yamamoto H, Lamphier MS, Fujita T, Taniguchi T, Harada H. The oncogenic transcription factor IRF-2 possesses a transcriptional repression and a latent activation domain. Oncogene 1994;9:1423-1428.

376. Xi H, Blanck G. The IRF-2 DNA binding domain facilitates the activation of the class II transactivator (CIITA) type IV promoter by IRF-1. Mol.Immunol. 2003;39:677-684.

377. Taniguchi T. IRF-1 and IRF-2 as regulators of the interferon system and cell growth. Indian J.Biochem.Biophys. 1995;32:235-239.

378. Nhu QM, Cuesta N, Vogel SN. Transcriptional regulation of lipopolysaccharide (LPS)-induced Toll-like receptor (TLR) expression in murine macrophages: role of interferon regulatory factors 1 (IRF-1) and 2 (IRF-2). J.Endotoxin.Res. 2006;12:285-295.

379. Watanabe N, Sakakibara J, Hovanessian AG, Taniguchi T, Fujita T. Activation of IFN-beta element by IRF-1 requires a posttranslational event in addition to IRF-1 synthesis. Nucleic Acids Res. 1991;19:4421-4428.

380. Harada H, Takahashi E, Itoh S et al. Structure and regulation of the human interferon regulatory factor 1 (IRF-1) and IRF-2 genes: implications for a gene network in the interferon system. Mol.Cell Biol. 1994;14:1500-1509.

381. Paun A, Pitha PM. The IRF family, revisited. Biochimie 2007;89:744-753.

382. Takaoka A, Yanai H, Kondo S et al. Integral role of IRF-5 in the gene induction programme activated by Toll-like receptors. Nature 2005;434:243-249.

383. Kaisho T, Tanaka T. Turning NF-kappaB and IRFs on and off in DC. Trends Immunol. 2008;29:329-336.

384. Orkin SH, Zon LI. Hematopoiesis: an evolving paradigm for stem cell biology. Cell 2008;132:631-644.

385. Gabriele L, Ozato K. The role of the interferon regulatory factor (IRF) family in dendritic cell development and function. Cytokine Growth Factor Rev. 2007;18:503-510.

386. Schiavoni G, Mattei F, Sestili P et al. ICSBP is essential for the development of mouse type I interferon-producing cells and for the generation and activation of CD8alpha(+) dendritic cells. J.Exp.Med. 2002;196:1415-1425.

387. Tsujimura H, Tamura T, Ozato K. Cutting edge: IFN consensus sequence binding protein/IFN regulatory factor 8 drives the development of type I IFN-producing plasmacytoid dendritic cells. J.Immunol. 2003;170:1131-1135.

388. Suzuki S, Honma K, Matsuyama T et al. Critical roles of interferon regulatory factor 4 in CD11bhighCD8alpha- dendritic cell development. Proc.Natl.Acad.Sci.U.S.A 2004;101:8981-8986.

389. Aliberti J, Schulz O, Pennington DJ et al. Essential role for ICSBP in the in vivo development of murine CD8alpha + dendritic cells. Blood 2003;101:305-310.

390. Tamura T, Tailor P, Yamaoka K et al. IFN regulatory factor-4 and -8 govern dendritic cell subset development and their functional diversity. J.Immunol. 2005;174:2573-2581.

391. Gabriele L, Fragale A, Borghi P et al. IRF-1 deficiency skews the differentiation of dendritic cells toward plasmacytoid and tolerogenic features. J.Leukoc.Biol. 2006;80:1500-1511.

392. Honda K, Mizutani T, Taniguchi T. Negative regulation of IFN-alpha/beta signaling by IFN regulatory factor 2 for homeostatic development of dendritic cells. Proc.Natl.Acad.Sci.U.S.A 2004;101:2416-2421.

393. Ichikawa E, Hida S, Omatsu Y et al. Defective development of splenic and epidermal CD4+ dendritic cells in mice deficient for IFN regulatory factor-2. Proc.Natl.Acad.Sci.U.S.A 2004;101:3909-3914.

394. Ahn JH, Lee Y, Jeon C et al. Identification of the genes differentially expressed in human dendritic cell subsets by cDNA subtraction and microarray analysis. Blood 2002;100:1742-1754.

395. Izaguirre A, Barnes BJ, Amrute S et al. Comparative analysis of IRF and IFN-alpha expression in human plasmacytoid and monocyte-derived dendritic cells. J.Leukoc.Biol. 2003;74:1125-1138.

396. Lehtonen A, Veckman V, Nikula T et al. Differential expression of IFN regulatory factor 4 gene in human monocyte-derived dendritic cells and macrophages. J.Immunol. 2005;175:6570-6579.

397. Negishi H, Fujita Y, Yanai H et al. Evidence for licensing of IFN-gamma-induced IFN regulatory factor 1 transcription factor by MyD88 in Toll-like receptor-dependent gene induction program. Proc.Natl.Acad.Sci.U.S.A 2006;103:15136-15141.

398. Negishi H, Ohba Y, Yanai H et al. Negative regulation of Toll-like-receptor signaling by IRF-4. Proc.Natl.Acad.Sci.U.S.A 2005;102:15989-15994.

399. Schmitz F, Heit A, Guggemoos S et al. Interferon-regulatory-factor 1 controls Toll-like receptor 9-mediated IFN-beta production in myeloid dendritic cells. Eur.J.Immunol. 2007;37:315-327.

400. Kawai T, Takeuchi O, Fujita T et al. Lipopolysaccharide stimulates the MyD88-independent pathway and results in activation of IFN-regulatory factor 3 and the expression of a subset of lipopolysaccharide-inducible genes. J.Immunol. 2001;167:5887-5894.

401. Doyle S, Vaidya S, O'Connell R et al. IRF3 mediates a TLR3/TLR4-specific antiviral gene program. Immunity. 2002;17:251-263.

402. Sakaguchi S, Negishi H, Asagiri M et al. Essential role of IRF-3 in lipopolysaccharide-induced interferon-beta gene expression and endotoxin shock. Biochem.Biophys.Res.Commun. 2003;306:860-866.

403. Fitzgerald KA, McWhirter SM, Faia KL et al. IKKepsilon and TBK1 are essential components of the IRF3 signaling pathway. Nat.Immunol. 2003;4:491-496.

404. Sharma S, tenOever BR, Grandvaux N et al. Triggering the interferon antiviral response through an IKK-related pathway. Science 2003;300:1148-1151.

405. Takaoka A, Yanai H. Interferon signalling network in innate defence. Cell Microbiol. 2006;8:907-922.

406. Honda K, Ohba Y, Yanai H et al. Spatiotemporal regulation of MyD88-IRF-7 signalling for robust type-I interferon induction. Nature 2005;434:1035-1040.

407. Uematsu S, Sato S, Yamamoto M et al. Interleukin-1 receptor-associated kinase-1 plays an essential role for Toll-like receptor (TLR)7- and TLR9-mediated interferon-{alpha} induction. J.Exp.Med. 2005;201:915-923.

408. Hoshino K, Sugiyama T, Matsumoto M et al. IkappaB kinase-alpha is critical for interferon-alpha production induced by Toll-like receptors 7 and 9. Nature 2006;440:949-953.

409. Guiducci C, Ghirelli C, Marloie-Provost MA et al. PI3K is critical for the nuclear translocation of IRF-7 and type I IFN production by human plasmacytoid predendritic cells in response to TLR activation. J.Exp.Med. 2008;205:315-322.

410. Kawai T, Sato S, Ishii KJ et al. Interferon-alpha induction through Toll-like receptors involves a direct interaction of IRF7 with MyD88 and TRAF6. Nat.Immunol. 2004;5:1061-1068.

411. Lin R, Yang L, Arguello M, Penafuerte C, Hiscott J. A CRM1-dependent nuclear export pathway is involved in the regulation of IRF-5 subcellular localization. J.Biol.Chem. 2005;280:3088-3095.

412. Cheng TF, Brzostek S, Ando O et al. Differential activation of IFN regulatory factor (IRF)-3 and IRF-5 transcription factors during viral infection. J.Immunol. 2006;176:7462-7470.

413. Yasuda K, Richez C, Maciaszek JW et al. Murine dendritic cell type I IFN production induced by human IgG-RNA immune complexes is IFN regulatory factor (IRF)5 and IRF7 dependent and is required for IL-6 production. J.Immunol. 2007;178:6876-6885.

414. Schoenemeyer A, Barnes BJ, Mancl ME et al. The interferon regulatory factor, IRF5, is a central mediator of toll-like receptor 7 signaling. J.Biol.Chem. 2005;280:17005-17012.

415. Gabriele L, Ozato K. The role of the interferon regulatory factor (IRF) family in dendritic cell development and function. Cytokine Growth Factor Rev. 2007;18:503-510.

416. Tsujimura H, Tamura T, Kong HJ et al. Toll-like receptor 9 signaling activates NF-kappaB through IFN regulatory factor-8/IFN consensus sequence binding protein in dendritic cells. J.Immunol. 2004;172:6820-6827.

417. Tailor P, Tamura T, Kong HJ et al. The feedback phase of type I interferon induction in dendritic cells requires interferon regulatory factor 8. Immunity. 2007;27:228-239.

418. Tsuno T, Mejido J, Zhao T et al. IRF9 is a key factor for eliciting the antiproliferative activity of IFN-alpha. J.Immunother. 2009;32:803-816.

419. Poat B, Hazari S, Chandra PK et al. Intracellular expression of IRF9 Stat fusion protein overcomes the defective Jak-Stat signaling and inhibits HCV RNA replication. Virol.J. 2010;7:265.

420. Pietila TE, Veckman V, Lehtonen A et al. Multiple NF-kappaB and IFN regulatory factor family transcription factors regulate CCL19 gene expression in human monocyte-derived dendritic cells. J.Immunol. 2007;178:253-261.

421. Takauji R, Iho S, Takatsuka H et al. CpG-DNA-induced IFN-alpha production involves p38 MAPK-dependent STAT1 phosphorylation in human plasmacytoid dendritic cell precursors. J.Leukoc.Biol. 2002;72:1011-1019.

422. Ilka Knippertz. Genetic and physical modification of human monocyte derived dendritic cells in order to improve vaccination protocols. 91-98. 1-7-2008. Ref Type: Thesis/Dissertation

423. Yamada N, Katz SI. Generation of mature dendritic cells from a CD14+ cell line (XS52) by IL-4, TNF-alpha, IL-1 beta, and agonistic anti-CD40 monoclonal antibody. J.Immunol. 1999;163:5331-5337.

424. Wolenski M, Cramer SO, Ehrlich S et al. Enhanced activation of CD83-positive T cells. Scand.J.Immunol. 2003;58:306-311.

425. Krausgruber T, Saliba D, Ryzhakov G et al. IRF5 is required for late-phase TNF secretion by human dendritic cells. Blood 2010;115:4421-4430.

426. Breloer M. CD83: regulator of central T cell maturation and peripheral immune response. Immunol.Lett. 2008;115:16-17.

427. Zinser E, Steinkasserer A. Published studies reporting the efficacy of soluble CD83 in vitro as well as in vivo. Immunol.Lett. 2008;115:18-19.

428. Fujimoto Y, Tedder TF. CD83: a regulatory molecule of the immune system with great potential for therapeutic application. J.Med.Dent.Sci. 2006;53:85-91.

429. Bartova E, Krejci J, Harnicarova A, Galiova G, Kozubek S. Histone modifications and nuclear architecture: a review. J.Histochem.Cytochem. 2008;56:711-721.

430. Cook PR. Nongenic transcription, gene regulation and action at a distance. J.Cell Sci. 2003;116:4483-4491.

431. Jaenisch R, Bird A. Epigenetic regulation of gene expression: how the genome integrates intrinsic and environmental signals. Nat.Genet. 2003;33 Suppl:245-254.

432. Felsenfeld G, Groudine M. Controlling the double helix. Nature 2003;421:448-453.

433. Peterson CL, Laniel MA. Histones and histone modifications. Curr.Biol. 2004;14:R546-R551.

434. Lam AL, Pazin DE, Sullivan BA. Control of gene expression and assembly of chromosomal subdomains by chromatin regulators with antagonistic functions. Chromosoma 2005;114:242-251.

435. Margueron R, Trojer P, Reinberg D. The key to development: interpreting the histone code? Curr.Opin.Genet.Dev. 2005;15:163-176.

436. Schubeler D, Francastel C, Cimbora DM et al. Nuclear localization and histone acetylation: a pathway for chromatin opening and transcriptional activation of the human beta-globin locus. Genes Dev. 2000;14:940-950.

437. Bulger M, Sawado T, Schubeler D, Groudine M. ChIPs of the beta-globin locus: unraveling gene regulation within an active domain. Curr.Opin.Genet.Dev. 2002;12:170-177.

438. Litt MD, Simpson M, Recillas-Targa F, Prioleau MN, Felsenfeld G. Transitions in histone acetylation reveal boundaries of three separately regulated neighboring loci. EMBO J. 2001;20:2224-2235.

439. Sen R, Oltz E. Genetic and epigenetic regulation of IgH gene assembly. Curr.Opin.Immunol. 2006;18:237-242.

440. Roh TY, Wei G, Farrell CM, Zhao K. Genome-wide prediction of conserved and nonconserved enhancers by histone acetylation patterns. Genome Res. 2007;17:74-81.

441. Nencioni A, Beck J, Werth D et al. Histone deacetylase inhibitors affect dendritic cell differentiation and immunogenicity. Clin.Cancer Res. 2007;13:3933-3941.

442. Drew PD, Franzoso G, Becker KG et al. NF kappa B and interferon regulatory factor 1 physically interact and synergistically induce major histocompatibility class I gene expression. J.Interferon Cytokine Res. 1995;15:1037-1045.

443. Kollet JI, Petro TM. IRF-1 and NF-kappaB p50/cRel bind to distinct regions of the proximal murine IL-12 p35 promoter during costimulation with IFN-gamma and LPS. Mol.Immunol. 2006;43:623-633.

444. Washizu J, Nishimura H, Nakamura N, Nimura Y, Yoshikai Y. The NF-kappaB binding site is essential for transcriptional activation of the IL-15 gene. Immunogenetics 1998;48:1-7.

445. Garoufalis E, Kwan I, Lin R et al. Viral induction of the human beta interferon promoter: modulation of transcription by NF-kappa B/rel proteins and interferon regulatory factors. J.Virol. 1994;68:4707-4715.

446. Neish AS, Read MA, Thanos D et al. Endothelial interferon regulatory factor 1 cooperates with NF-kappa B as a transcriptional activator of vascular cell adhesion molecule 1. Mol.Cell Biol. 1995;15:2558-2569.

447. Feuillard J, Gouy H, Bismuth G et al. NF-kappa B activation by tumor necrosis factor alpha in the Jurkat T cell line is independent of protein kinase A, protein kinase C, and Ca(2+)-regulated kinases. Cytokine 1991;3:257-265.

448. Watanabe S, Yssel H, Harada Y, Arai K. Effects of prostaglandin E2 on Th0-type human T cell clones: modulation of functions of nuclear proteins involved in cytokine production. Int.Immunol. 1994;6:523-532.

449. Baker RG, Hayden MS, Ghosh S. NF-kappaB, inflammation, and metabolic disease. Cell Metab 2011;13:11-22.

450. Rescigno M, Martino M, Sutherland CL, Gold MR, Ricciardi-Castagnoli P. Dendritic cell survival and maturation are regulated by different signaling pathways. J.Exp.Med. 1998;188:2175-2180.

451. Ouaaz F, Arron J, Zheng Y, Choi Y, Beg AA. Dendritic cell development and survival require distinct NF-kappaB subunits. Immunity. 2002;16:257-270.

452. Boffa DJ, Feng B, Sharma V et al. Selective loss of c-Rel compromises dendritic cell activation of T lymphocytes. Cell Immunol. 2003;222:105-115.

453. Grumont R, Hochrein H, O'Keeffe M et al. c-Rel regulates interleukin 12 p70 expression in CD8(+) dendritic cells by specifically inducing p35 gene transcription. J.Exp.Med. 2001;194:1021-1032.

454. Carmody RJ, Ruan Q, Liou HC, Chen YH. Essential roles of c-Rel in TLR-induced IL-23 p19 gene expression in dendritic cells. J.Immunol. 2007;178:186-191.

455. Lu YC, Kim I, Lye E et al. Differential role for c-Rel and C/EBPbeta/delta in TLR-mediated induction of proinflammatory cytokines. J.Immunol. 2009;182:7212-7221.

456. Fujita T, Nolan GP, Ghosh S, Baltimore D. Independent modes of transcriptional activation by the p50 and p65 subunits of NF-kappa B. Genes Dev. 1992;6:775-787.

457. Taniguchi T. Transcription factors IRF-1 and IRF-2: linking the immune responses and tumor suppression. J.Cell Physiol 1997;173:128-130.

458. Matsuyama T, Kimura T, Kitagawa M et al. Targeted disruption of IRF-1 or IRF-2 results in abnormal type I IFN gene induction and aberrant lymphocyte development. Cell 1993;75:83-97.

459. Zhang S, Thomas K, Blanco JC, Salkowski CA, Vogel SN. The role of the interferon regulatory factors, IRF-1 and IRF-2, in LPS-induced cyclooxygenase-2 (COX-2) expression in vivo and in vitro. J.Endotoxin.Res. 2002;8:379-388.

460. Salkowski CA, Kopydlowski K, Blanco J et al. IL-12 is dysregulated in macrophages from IRF-1 and IRF-2 knockout mice. J.Immunol. 1999;163:1529-1536.

461. Hu Y, Park-Min KH, Yarilina A, Ivashkiv LB. Regulation of STAT pathways and IRF1 during human dendritic cell maturation by TNF-alpha and PGE2. J.Leukoc.Biol. 2008;84:1353-1360.

462. Wathelet MG, Lin CH, Parekh BS et al. Virus infection induces the assembly of coordinately activated transcription factors on the IFN-beta enhancer in vivo. Mol.Cell 1998;1:507-518.

463. Kim TK, Maniatis T. The mechanism of transcriptional synergy of an in vitro assembled interferon-beta enhanceosome. Mol.Cell 1997;1:119-129.

464. Sgarbanti M, Remoli AL, Marsili G et al. IRF-1 is required for full NF-kappaB transcriptional activity at the human immunodeficiency virus type 1 long terminal repeat enhancer. J.Virol. 2008;82:3632-3641.

465. Masumi A. Histone acetyltransferases as regulators of nonhistone proteins: the role of interferon regulatory factor acetylation on gene transcription. J.Biomed.Biotechnol. 2011;2011:640610.

466. Masumi A, Ozato K. Coactivator p300 acetylates the interferon regulatory factor-2 in U937 cells following phorbol ester treatment. J.Biol.Chem 2001;276:20973-20980.

467. Marsili G, Remoli AL, Sgarbanti M, Battistini A. Role of acetylases and deacetylase inhibitors in IRF-1-mediated HIV-1 long terminal repeat transcription. Ann.N.Y.Acad.Sci. 2004;1030:636-643.

468. Mancl ME, Hu G, Sangster-Guity N et al. Two discrete promoters regulate the alternatively spliced human interferon regulatory factor-5 isoforms. Multiple isoforms with distinct cell type-specific expression, localization, regulation, and function. J.Biol.Chem. 2005;280:21078-21090.

References

469. Barnes BJ, Kellum MJ, Field AE, Pitha PM. Multiple regulatory domains of IRF-5 control activation, cellular localization, and induction of chemokines that mediate recruitment of T lymphocytes. Mol.Cell Biol. 2002;22:5721-5740.

470. Cheng TF, Brzostek S, Ando O et al. Differential activation of IFN regulatory factor (IRF)-3 and IRF-5 transcription factors during viral infection. J.Immunol. 2006;176:7462-7470.

471. Krausgruber T, Blazek K, Smallie T et al. IRF5 promotes inflammatory macrophage polarization and TH1-TH17 responses. Nat.Immunol. 2011;12:231-238.

472. Chae M, Kim K, Park SM et al. IRF-2 regulates NF-kappaB activity by modulating the subcellular localization of NF-kappaB. Biochem.Biophys.Res.Commun. 2008;370:519-524.

Acknowledgements

- Zunächst möchte ich mich bei Prof. Dr. G. Schuler dafür bedanken, dass er mich so freundlich an seiner Klinik aufgenommen hat.

- Mein Dank gilt auch Herrn Prof. Dr. L. Nitschke, da er diese Arbeit stellvertretend für den Fachbereich Biologie der Naturwissenschaftlichen Fakultät betreut hat. Ebenfalls möchte ich mich bei Prof. Dr. E. Kämpgen bedanken, der die Betreuung von Seiten der Hautklinik übernommen hat. Ferner möchte ich Prof. Dr. F. Nimmerjahn danken, der den Vorsitz meiner Dissertationsprüfung übernommen hat.

- Ganz herzlich möchte ich mich bei Prof. Dr. A. Steinkasserer bedanken, der mir nicht nur die materiellen Möglichkeiten zur Verfügung gestellt hat, in seiner Arbeitsgruppe zu promovieren, sondern mich auch hervorragend während meiner Labortätigkeit betreut und unterstützt hat.

- Ferner danke ich Prof. Dr. B. Fleckenstein für die freundliche Aufnahme in das Graduiertenkolleg 1071. Großer Dank gilt auch Dr. B. Biesinger für die kompetente Beratung und Betreuung im Rahmen des GRK. An dieser Stelle möchte ich auch allen anderen Mitgliedern des GRK für die Unterstützung während meiner Promotionszeit danken. Für die großartige Unterstützung bei der Durchführung der EMSAs, sowohl durch Bereitstellung der Laborräume, als auch durch Weitergabe ihres Know Hows, möchte ich mich bei der ganzen AG Stürzl, insbesondere bei Dr. L. Naschberger und Prof. Dr. M. Stürzl, bedanken. Ferner gilt mein Dank noch Prof. Dr. T. Stamminger, der mich im Vorfeld hervorragend in Bezug auf die EMSAs beraten hat.

- Ein ganz besonderer Dank gilt Dr. I. Knippertz, die die Urheberin dieses Projektes ist und mich „mit ins Boot geholt" hat. Sie stand mir als Betreuerin und gute Kollegin stets mit Rat und Tat zur Seite.

- Ebenfalls besonders badanken möchte ich mich bei Dr. T. Werner, der uns mit seiner Expertise und fantastischen bioinformatioschen Analyse der regulatorischen Elemente am CD83 Genlokus unterstützt hat.

- K. Blume und A. Deinzer möchte ich für die liebe Unterstützung bei vielen praktischen Arbeiten im Labor und für das Vermitteln von Wissen und Techniken danken. Sehr dankbar bin ich auch für das hervorragende Arbeitsklima an der Hautklinik: Alle Mitglieder der „Schuppen"- und „Container"-Crew sowie der „RNA-Gruppe" haben mir stets mit freundlichen Worten und schönen Aktivitäten auch außerhalb des Labors Mut gemacht.

- Ferner gilt auch meinen Eltern ein besonderer Dank, da diese mir mein Studium und damit auch meine Promotion ermöglicht haben. Meinen beiden Schwestern Miriam und Ilaria möchte ich in diesem Zusammenhang besonders für den Rückhalt während den schwierigen Phasen der Promotion danken.

- Nonno e nonna, grazie per l'incoraggiamento durante i tempi stressanti.

- Ganz herzlicher Dank geht an meine Freunde P. Luff und C. Reich, die mich seit Beginn meines Studiums begleiten und mich in jeder schwierigen Lage mit Rat und Tat unterstützt haben. Auch meinem guten Freund, J. Hempfling, möchte ich an dieser Stelle ganz besonders danken, da er vor allem in den anstrengenden Phasen meiner Promotion für die nötige Ablenkung gesorgt hat.

- In ganz besonderem Maße möchte ich meiner Ehefrau Dr. R. Jochmann danken. Sie hat mir im privaten, wie auch im wissenschaftlichen Bereich den Rücken gestärkt hat. Ihre Expertise in experimentellen Belangen hat mir an vielen Wegpunkten weitergeholfen und ihre liebevolle Unterstützung war mir die wertvollste Motivation während der schwierigeren Zeiten meiner Promotion.

Patents, Publications, Presentations, Participations and Commitments (as of January 2013)

Patents

- Knippertz,I., **Stein,M. F.**, Steinkasserer,A., Werner,T. (2011). Characterization of functional human dendritic cell-specific CD83 promoter/enhancer regions and the use thereof for the treatment or prevention of diseases or medical conditions related to malignancy, autoimmunity or prevention of transplant rejections. European Patent Office, application number EP11164344. Receipt acknowledged April 29th, 2011

Publications

- **Stein,M.F.**, Lang,S., Winkler,T., Deinzer,A., Erber,S., Nettelbeck,D.M., Naschberger,E., Jochmann,R., Stürzl,M., Slany RK., Werner,T., Steinkasserer,A., Knippertz,I. (2013). Multiple IRF- and NFkB-sites cooperate in mediating cell type- and maturation-specific activation of the human CD83 promoter in dendritic cells. MCB. MCB Accepts, published online ahead of print on 22 January 2013. Mol. Cell. Biol. doi:10.1128/MCB.01051-12.

- Knippertz,I., **Stein,M.F.**, Dörrie,J., Schaft,N., Müller,I., Deinzer,A., Steinkasserer,A., Nettelbeck,D.M., (2011). *Mild hyperthermia enhances human monocyte-derived dendritic cell functions and offers potential for applications in vaccination strategies.* International Journal of Hyperthermia, Sept. 2011; 27(6): 591-603.

Talks

- *Characterization of the human CD83 promoter reveals a complex NFkB/IRF-framework governing expression in mature dendritic cells.* Meeting of the European Macrophage and Dendritic Cell Society (EMDS), September 2012, Debrecen, Hungary.

- *The Regulation of CD83.* 7th Retreat of the Graduiertenkolleg Viruses of the Immune System (GRK1071), April 2009, Schlaifhausen, Germany.

- *The CD83 promoter - a tough nut to crack.* 1st Think Tank Symposium Dermatology Erlangen, November 2009, Heiligenstadt, Germany.

- *A Closer Look Upon CD83 Regulation.* 6th Retreat of the Graduiertenkolleg Viruses of the Immune System (GRK1071), September 2008, New England Primate Research Center (NEPRC) of Harvard Medical School in Southborough, Massachusetts, USA.

- *CD83 Is Not Just A Maturation Marker On Dendritic Cells: Characterizing CD83.* 5th Retreat of the Graduiertenkolleg Viruses of the Immune System (GRK1071), April 2008, Schlaifhausen, Germany.

- *CD83 Signaling - A Basic Approach.* 4th Retreat of the Graduiertenkolleg Viruses of the Immune System (GRK1071), April 2007, Schlaifhausen, Germany.

Poster presentations

- *Characterization of the human CD83 promoter reveals a complex NFkB/IRF-framework governing expression in mature dendritic cells.* 12th International Symposium on Dendritic Cells (DC2012), October 2012, Daegu, Korea.

- *Characterization of the human CD83 promoter reveals a complex NFkB/IRF-framework governing expression in mature dendritic cells.* Meeting of the European Macrophage and Dendritic Cell Society (EMDS), September 2012, Debrecen, Hungary.

- *Tamiflu and Relenza – Im Kampf gegen die Grippe.* Lange Nacht der Wissenschaften 2009, October 2009, Erlangen, Germany.

- *A closer look upon CD83 Regulation.* 2nd International GRK-Symposium "Regulators of Adaptive Immunity", October 2008, Erlangen, Germany.

- *Tamiflu – Rettung vor der Vogelgrippe?* Lange Nacht der Wissenschaften 2007, October 2007, Erlangen, Germany.

Commitments

- Commitee *"Public Relations"*, Graduiertenkolleg Viruses of the Immune System (GRK1071), August 2006- August 2009, Erlangen, Germany.

- Chairman of session VI International GRK Symposium Introduction. 2nd International GRK-Symposium *"Regulators of Adaptive Immunity"*, October 2008, Erlangen, Germany.

Workshops, symposia and congresses

- Participation and presentation at the 12th International Symposium on Dendritic Cells (DC2012), October 2012, Daegu, Korea. Poster.

- Participation and presentation at the 2012 Meeting of the European Macrophage and Dendritic Cell Society (EMDS), September 2012, Debrecen, Hungary. Poster and talk.

- Participation at the Fourth Weißenburg Symposium "Epigenetics and the Control of Gene Expression", June 2011, Weißenburg, Germany.

- Participation at the "Cellular Therapy International" Symposium 2011, March 2011, Erlangen, Germany.

- Participation and presentation at the 1st "Think Tank" Symposium Dermatology Erlangen, November 2009, Heiligenstadt, Germany.

- Participation and presentation at the **7th** Retreat of the Graduiertenkolleg Viruses of the Immune System (GRK1071), April 2009, Schlaifhausen, Germany.

- Participation at the soft skill course "Preparing and Presenting a Poster Professionally" in English, March 2009, Erlangen, Germany.

- Working technique courses "FACS, 2-D DIGE, Immune-histology, RT-PCR, Bioinformatics, EDP-Tools, PowerPoint", 2008/09, Erlangen, Germany.

- Participation and chairman at the 2nd International GK Symposium, "Regulators of Adaptive Immunity", October 2008, Erlangen, Germany.

- Participation and presentation at the **6th** Retreat of the Graduiertenkolleg Viruses of the Immune System (GRK1071), September 2008, New England Primate Research Center (NEPRC) of Harvard Medical School in Southborough, Massachusetts, USA.

- Participation at the soft skill course "Presenting Professionally" in English, August 2008, Erlangen, Germany.

- Participation and presentation at the 5th Retreat of the Graduiertenkolleg Viruses of the Immune System, April 2008, Schlaifhausen, Germany.

- Participation at the jobseminar "Promotion an der FAU - und dann?", January 2008, Erlangen, Germany.

- Participation at the 11th Joint Meeting Signal Transduction Society "Signal Transduction, Receptors, Mediators and Genes", November 2007, Weimar, Germany.

- Participation at the 5th International Meeting on Dendritic Cell Vaccination and other Strategies to Tip the Balance of the Immune System (DC2007). July 2007, Bamberg, Germany.

- Participation and presentation at the 4th Retreat of the Graduiertenkolleg Viruses of the Immune System, April 2007, Schlaifhausen, Germany
- Participation at the soft skill course "Effective Scientific Writing" in English, January 2007, Erlangen, Germany.

- Participation at the 1st International GK Symposium, "Regulators of Adaptive Immunity", September 2006, Erlangen, Germany.

- Participation at the workshop "Forschungsmethodik", reviewing evaluation of statistical data, November 2006, Erlangen, Germany

Prizes

- PhD sholarship 2006 – 2009. Promotionsstipendium der Deutschen Forschungsgemeinschaft innerhalb des Graduiertenkollegs 1071 *Viren des Immunsystems*.

i want morebooks!

Buy your books fast and straightforward online - at one of world's fastest growing online book stores! Environmentally sound due to Print-on-Demand technologies.

Buy your books online at
www.get-morebooks.com

Kaufen Sie Ihre Bücher schnell und unkompliziert online – auf einer der am schnellsten wachsenden Buchhandelsplattformen weltweit! Dank Print-On-Demand umwelt- und ressourcenschonend produziert.

Bücher schneller online kaufen
www.morebooks.de

VDM Verlagsservicegesellschaft mbH
Heinrich-Böcking-Str. 6-8 Telefon: +49 681 3720 174 info@vdm-vsg.de
D - 66121 Saarbrücken Telefax: +49 681 3720 1749 www.vdm-vsg.de

Printed by Books on Demand GmbH, Norderstedt / Germany